SENSORS AND ACTUATORS

SENSORS AND ACTUATORS

Proceedings of a Conference held at the
Manchester Business School
15–16 July 1996

Edited by

D.A. Hall and C.E. Millar

Book 714
Published in 1999 by
IOM Communications Ltd
1 Carlton House Terrace
London SW1Y 5DB

IOM Communications Ltd
is a wholly-owned subsidiary of
the Institute of Materials

ISBN 1 86125 089 4

Typeset by IOM Communications Ltd

Printed and bound in the UK by
The University Press, Cambridge

CONTENTS

vi   *Contents*

FOREWORD

The IOM conference on Sensors and Actuators was held at Manchester Business School, University of Manchester, from 15–16 July, 1996. The meeting was organised on behalf of the Electronic Applications Divisional Board (formerly the Electrical, Magnetic and Optical Materials Committee) of the Institute of Materials. There were 40 delegates to the conference, from various UK institutions and industries. The programme for the conference comprised 22 oral presentations in 6 sessions, and 14 posters.

Our overwhelming impression from the conference was of the diverse nature of the subject, both in terms of the physical phenomena which can be exploited in electromechanical sensors and actuators and the varied backgrounds of the research personnel. It seems inevitable that the most significant advances in this field will come from research teams involving collaborations across the traditional boundaries between science and engineering subjects and close cooperation between industry and academia. There are many opportunities for collaborative research in this area; the main barrier is perhaps our lack of awareness of what each of us has to offer to the subject.

The structure of the Proceedings is along the same lines as the oral programme during the conference, with the papers being classified according to the underlying physical phenomena responsible for the sensor/actuator function and/or the type of device being investigated. The topics covered range from fundamental materials studies to the design and fabrication of prototype devices. Therefore, this collection of papers reflects the broad range of interest in the subject and represents a cross section of the much larger UK activity in sensors and actuators.

Finally, we would like to thank all of the authors for their contributions and apologise for the long delay in preparing the papers for publication. We feel that the Proceedings still represents a useful contribution to the subject and hope that it might help to encourage further collaboration within the UK Science and Engineering community.

D.A. Hall and C.E. Millar

DISTRIBUTED PVDF SENSORS FOR ACTIVE STRUCTURAL ACOUSTIC CONTROL

M. E. JOHNSON*, S. J. ELLIOTT AND M. J. BRENNAN

*Vibration and Acoustics Laboratories, Virginia Polytechnic Institute and State University, Dept. of Mechanical Engineering, College of Engineering, Blacksburg, Virginia 24061, USA

Institute of Sound and Vibration Research, University of Southampton, Highfield, Southampton SO17 1BJ

ABSTRACT

The acoustic radiation from a vibrating structure can be reduced by an active control system provided the system has a suitable sensing capability. The radiation of sound from a structure is in general a spatially distributed phenomenon and it is often necessary to extract information about the behaviour of the structure over its entire surface. The major source of acoustic radiation from a structure at low frequencies is the integrated normal velocity or *volume velocity* of the structure. For the purposes of active control it is therefore important that this quantity be measured. This can be achieved by using a number of point sensors, but the complexity of the system required to deal with these sensors can become prohibitive. Alternatively, a single distributed PVDF sensor can be used to measure the volume velocity of a structure due to bending, but unfortunately such a sensor is incapable of detecting the whole body motion of the structure. An alternative distributed sensor which uses PVDF as a distributed accelerometer (in the d_{33} mode), can in principle, directly measure the volume velocity of the structure due to any arbitrary motion. Unfortunately, PVDF is very sensitive to strain on the structure's surface and this masks any signal output due to acceleration. Methods of desensitising the PVDF sensor to surface strain are discussed.

INTRODUCTION

The active control of sound radiation from structures has in recent years become a popular topic for research. One of the main areas of interest has been the development of structurally mounted sensors and actuators which are suitable for the control the sound radiation. Structural transducers are often preferred to acoustic transducers for a number of reasons, (i) structural actuators are often lighter than acoustic sources, (ii) the complexity of the control system required to achieve good control can be reduced and (iii) time delays between structural sensors and actuators can be smaller than the time delays between acoustic sensors and actuators.

Although the vibration of a surface may be very complex, the main mechanism of sound radiation at low frequencies is relatively simple.[1,2] If the acoustic wavelength in air is large compared to the size of the structure it is the volume velocity or integrated normal velocity of the structure that accounts for most of the sound radiation.[2] Using a control actuator to minimise or cancel the output of a volume velocity sensor will therefore reduce the sound power radiation from a structure at low frequencies. The development of volume velocity sensors has therefore been the subject of a number of publications.[3-5] For most of this paper the sensors developed will measure volume displacement which is the integral of volume velocity with respect to time. Therefore, volume displacement sensors coupled with a differentiator can be used to measure volume velocity.

Although volume velocity is a single parameter it is dependent on the behaviour of the entire surface and to measure this quantity accurately requires a spatially distributed sensor or a number of discrete sensors that are spatially distributed.

DISTRIBUTED SENSORS

Spatially distributed sensors produce outputs which are a function of the response of the system over an area and are therefore integrating transducers. For the purposes of controlling the low frequency acoustic radiation from a surface the ideal distributed sensor would measure the integral of the surface displacement normal to the surface.

$$U = \int_A w(x, y) \mathrm{d}A \tag{1}$$

where U is the integrated displacement or volume displacement of the surface A, which is in the x–y plane and whose normal surface displacement (i.e. in the z direction) is defined by w which is a function of position x and y. The displacement is also a function of time t but this will be suppressed for most of the equations in this paper. The volume velocity of the surface is given by $\partial U / \partial t$.

DISPLACEMENT SENSORS WHICH MEASURE SURFACE STRAIN

A piezoelectric material such as PVDF, when attached to the surface of a structure (Fig. 1), will produce a charge output q which is given by,[6]

$$q = \int_A D_3 \, \mathrm{d}A = \int_A \varepsilon_{33} E_3 + e_{31} S_1 + e_{32} S_2 + e_{36} S_6 \, \mathrm{d}A \tag{2}$$

where D_3 is the electric charge displacement per unit area on the surface electrode of the PVDF, ε is the permitivity, E_3 is the electric field intensity across the 3-axis of the film e_{31}, e_{32} and e_{36} are piezoelectric constants and S_1, S_2 and S_6 are the strain along the 1 and 2 axes

Orientation of plate and piezoelectric

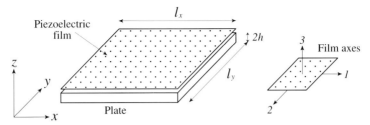

Fig. 1   The orientation and axes for the plate and the piezoelectric film.

and the shear strain in the 1–2 plane. The strains are all functions of position on the surface A. Therefore the output of the sensor in this case integrates the weighted sum of the strains over the surface.

   Throughout this paper the 1,2 and 3 axes of the PVDF will correspond to the x, y and z axes of the structure which will be taken to be a rectangular isotropic plate of dimensions l_x, l_y and thickness $2h$, as illustrated in Fig. 1.

   If a low impedance charge amplifier is used to measure the charge output of the film the value of E_3 will be set to zero. Also, most commercially available PVDF is quoted as having an e_{36} value of zero.[7] A PVDF sensor which covers the entire surface of the plate will therefore produce a charge output given by,

$$q = \int_0^{l_y} \int_0^{l_x} [e_{31}S_1 + e_{32}S_2]\,dxdy \tag{3}$$

To create a sensor which measures some desired distributed quantity the sensitivity of the sensor must be able to vary spatially. If this spatial sensitivity is described by the function $\Lambda(x, y)$ then the charge output will be given by,

$$q = \int_0^{l_y} \int_0^{l_x} \Lambda(x,y)[e_{31}S_1 + e_{32}S_2]\,dxdy \tag{4}$$

If the stretching of the midplane of the plate is ignored eqn (4) can be written in terms of the surface displacement of the plate $\omega(x, y)$,[6]

$$q = \int_0^{l_y} \int_0^{l_x} -h\Lambda(x,y)\left[e_{31}\frac{\partial^2 w(x,y)}{\partial x^2} + e_{32}\frac{\partial^2 w(x,y)}{\partial y^2} \right] dxdy \tag{5}$$

where h is half the thickness of the plate which corresponds to the film-neutral axis separation. If the above equation is integrated twice by parts the charge output can be re-written as

$$
q = he_{31} \int_0^{l_y} \left[-\left[\Lambda \frac{\partial w(x,y)}{\partial x} \right]_0^{l_x} + \left[\frac{\partial \Lambda}{\partial x} w(x,y) \right]_0^{l_x} - \int_0^{l_x} \frac{\partial^2 \Lambda}{\partial x^2} w(x,y) dx \right] dy
$$

$$
+ he_{32} \int_0^{l_x} \left[-\left[\Lambda \frac{\partial w(x,y)}{\partial x} \right]_0^{l_y} + \left[\frac{\partial \Lambda}{\partial y} w(x,y) \right]_0^{l_y} - \int_0^{l_y} \frac{\partial^2 \Lambda}{\partial y^2} w(x,y) dx \right] dx
$$

(6)

Depending on the boundary conditions of the plate some of the terms in the above equation can be set to zero. If, for instance, the plate is clamped then w, $\partial w/\partial x$ and $\partial w/\partial y$ are zero at the boundaries and the first, second, fourth and fifth terms in eqn (6) equal zero. The sensitivity function Λ can also be chosen such that it is equal to zero at the boundaries. For example, the sensitivity can be chosen to be quadratic in the x-direction and constant in the y-direction,[8]

$$
\Lambda = \alpha(l_x x - x^2)
$$

(7)

The sensitivity Λ is assumed to have a maximum value of one which corresponds to $\alpha = 4/l_x^2$. If the plate is clamped, then eqn (6) reduces to,

$$
q = 2he_{31} \alpha \int_0^{l_y} \int_0^{l_x} w(x,y) dx dy = 2he_{31} \alpha U
$$

(8)

In this case the output is proportional to the integrated displacement or volume displacement of the surface U, given in eqn (1). If a control actuator was used to cancel the output of such a sensor then the sound power radiation from the plate at low frequencies would be greatly reduced. Volume velocity sensors can also be designed for plates with pinned boundary conditions but involve the use of two PVDF films.[9]

Implementing a spatial sensitivity

A simple method of varying the sensitivity of a PVDF sensor spatially is to cut or etch shapes into the surface electrode of the film. To approximate the quadratic sensitivity given by eqn (7) a number of strips of quadratically varying width were etched into the surface of a large sheet of PVDF film (Fig. 2).

This sensor was fixed to a clamped plate and the differentiated output compared, when the plate was excited, with the sum of the outputs of a 7 by 7 array of point sensors which

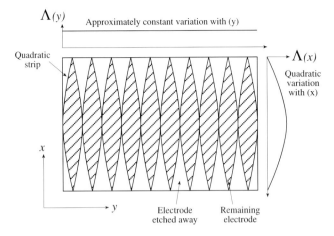

Fig. 2 A number of quadratically shaped strips etched into the electrode of a piece of piezoelectric film.

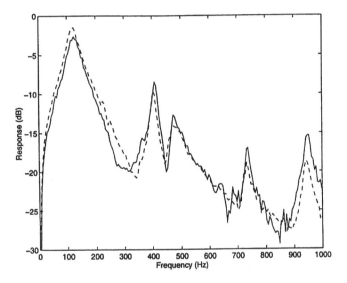

Fig. 3 The volume velocity estimated using 49 laser vibrometer measurements (dashed) and the output of the piezofilm sensor (solid) when the plate was excited using an acoustic source.

measured velocity. Figure 3 shows a comparison between the outputs of the sensor and the sum of the outputs of 49 laser vibrometer measurements, over a range of frequencies, when the plate is excited by an acoustic source.[4]

Figure 4 shows the results of another experiment where the differentiated output of the sensor is compared with the sum of the outputs of 49 point measurements when the plate is excited using a structural actuator.

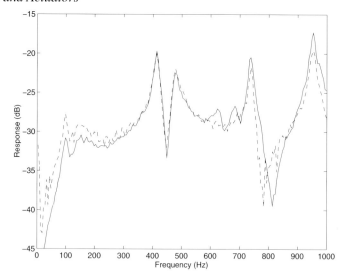

Fig. 4   The volume velocity estimated using 49 laser vibrometer measurements (dashed) and the output of the piezofilm sensor (solid) when the plate was excited using a structural source.

These results show good agreement between the output of the sensor and the volume displacement estimated using a number of point sensors. It also demonstrates the potential reduction of the number of sensors required from forty nine to one. It was found experimentally that by actively reducing the output of this sensor, using a secondary structural actuator, large reductions in the sound power radiation from the plate were achieved at low frequencies when the plate was excited by a primary acoustic source.

Disadvantages of using surface strain sensors

The main problem with the use of surface strain sensors for estimating the displacement of a surface is that, (i) the sensor cannot detect the whole body displacement of the surface (i.e. the first three modes of a free-free plate) and (ii) the sensor design must be tailored to suit the structure.

The strain at the surface is proportional to the second derivative of the normal displacement (eqn (5)) and therefore does not contain all of the information required to reconstruct the displacement $w(x, y)$ since differentiation disregards the d.c. component of a function. To reconstruct the displacement additional information is required in the form of boundary conditions. In any real application additional calibrated sensors may be required to measure the surface displacement of a surface vibrating with arbitrary motion.

The relationship between strain and displacement may also vary depending on the geometry of the structure. If there are discontinuities or changes in stiffness across the

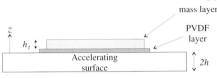

Fig. 5   PVDF film used as an accelerometer.

surface of the structure these factors must be taken into account. This implies that the design of surface strain sensors to measure volume velocity must be tailored to each particular application.

DISPLACEMENT SENSORS WHICH MEASURE ACCELERATION

Alternatively, the volume velocity of a surface may be measured using the piezoelectric material in the 3-direction. For example, most accelerometers use an inertial mass to produce a force on a piece of piezoelectric material when undergoing an acceleration. A distributed accelerometer could be created using a layer of PVDF film with a proof mass layer attached to its surface (Fig. (5)) and would not require any variation in spatial sensitivity (i.e. $\Lambda=1$).

The sensor output due to this acceleration component q_{acc} for a rectangular plate of dimensions l_x and l_y is given by,

$$q_{acc} = \rho h_t d_{33} \int_0^{l_y} \int_0^{l_x} \frac{\partial^2 w(x,y,t)}{\partial t^2} \, dxdy \qquad (9)$$

where ρ is the density of the proof mass layer, h_t is the thickness of the proof mass layer and d_{33} is the piezoelectric constant relating the charge per unit area on the electrode due to a stress in the 3-direction. For a given frequency of vibration ω the output can be written as,

$$q_{acc} = \rho h_t d_{33} \omega^2 \int_0^{l_y} \int_0^{l_x} w(x,y) \, dxdy \qquad (10)$$

The output is proportional to the integrated displacement or volume displacement of the surface. This result would be possible for any arbitrary motion including any whole body motion of the plate. Also, because the required spatial sensitivity is uniform the design of such a sensor would not have to take into account any changes or discontinuities in the structure.

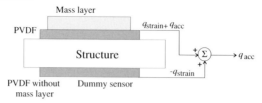

Fig. 6   Minimising the sensitivity to surface strain by combining the output of the sensor with the output of a 'dummy' sensor.

Sensitivity to strain

The sensor described above also has a sensitivity to surface strain which is given by eqn (5) with $\Lambda = 1$. The total output of the sensor will therefore be given by $q_{tot} = q_{strain} + q_{acc}$ where q_{strain} is the output due to strain (eqn (5)). It may be possible to compensate for the strain component q_{strain} by using a 'dummy' sensor which senses the surface strain but does not have a proof mass layer to provide the acceleration component. Figure 6 shows an example of how such a sensor could be made. If the two sensors were placed on either side of the structure then they would both experience similar surface strain (with opposite sign) as long as the proof mass layer had low stiffness in comparison to the structure. The output of these two sensors could then be summed to produce the output q_{acc}. This method however will only be successful in practice if the strain component q_{strain} is similar or smaller in magnitude than the acceleration component q_{acc}.

To calculate the relative magnitudes of these components a simple example will be taken. Consider a square simply supported aluminium plate of length l covered with a layer of biaxial PVDF film (i.e. $e_{31} = e_{32}$). For a simply supported plate the displacement can be approximated using a summation of modes.

$$w(x,y) = \sum_{n=1}^{\infty} \sum_{m=1}^{\infty} a_{nm} \sin(n\pi x / l)\sin(m\pi y / l) \tag{11}$$

where a_{nm} is the amplitude of the nm^{th} mode and l is the length of the square plate. In this case eqn (5) becomes,

$$q_{strain} = 4he_{31} \sum_{n=1}^{\infty} \sum_{m=1}^{\infty} a_{nm} \left[\frac{n}{m} + \frac{m}{n} \right] \tag{12}$$

since $e_{31} = e_{32}$ and $l_x = l_y$. The summation is only taken when both n and m are odd. For this example the acceleration component of the output q_{acc} (eqn (10)) becomes,

$$q_{acc} = \frac{4\rho h_t d_{33} l^2 \omega^2}{\pi^2} \sum_{n=1}^{\infty} \sum_{m=1}^{\infty} \frac{a_{nm}}{nm} \tag{13}$$

Again the summation is only taken when both n and m are odd. The ratio of these two outputs for the nm^{th} mode only is given by,

$$\left[\frac{q_{acc}}{q_{strain}}\right] = -\frac{\rho h_t d_{33}\omega^2}{e_{31}h} \frac{1}{\left[\left(\frac{n\pi}{l}\right)^2 + \left(\frac{m\pi}{l}\right)^2\right]} \tag{14}$$

For a simply supported plate the natural frequency of the nm^{th} mode (ω_{nm}) is given by,[10]

$$\omega_{mn} = \left[\frac{Eh^2}{3\rho_p(1-v^2)}\right]^{\frac{1}{2}}\left[\left(\frac{n\pi}{l}\right)^2 + \left(\frac{m\pi}{l}\right)^2\right] \tag{15}$$

where ρ_p is the plate's density, E its Young's modulus and v its Poisson's ratio. For a single odd-odd mode nm the ratio of sensitivities can be expressed in terms of the natural frequency ω_{nm} by substituting eqn (15) into eqn (14). The piezoelectric constant e_{31} can also be expressed as a function of the piezoelectric constants d_{31} and d_{32}[6] and for this example $d_{31} = d_{32}$ so the relationship can be simplified to $e_{31} = d_{31}E/(1-v)$.

$$\left[\frac{q_{acc}}{q_{strain}}\right]_{nm} = -\frac{(1-v)\,\rho h_t d_{33}\omega^2}{d_{31}\left[3\rho_p E\,(1-v^2)\right]^{\frac{1}{2}}\omega_{nm}} \tag{16}$$

If it is assumed that the mode is operating at resonance, where the contribution from the mode is the largest ω is equal to ω_{nm} and the above equation becomes,

$$\left[\frac{q_{acc}}{q_{strain}}\right]_{nm} = -\frac{(1-v)\,\rho h_t d_{33}\omega_{nm}}{d_{31}\left[3\rho_p E\,(1-v^2)\right]^{\frac{1}{2}}} \tag{17}$$

For aluminium $\omega_p = 2700$ kgm^{-3}, $v = 0.3$ and $E = 6.9$ x 10^{10} Nm^{-2}. A proof mass layer of 1kg per square meter (i.e. $\rho h_t = 1$) is assumed. Substituting these values into eqn (17) gives,

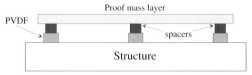

Fig. 7   The mass layer raised above the PVDF on a series of compliant spacers to increase the mass to area ratio.

$$\left[\frac{q_{acc}}{q_{strain}} \right]_{nm} = -3.1 \times 10^{-8} \frac{d_{33}}{d_{31}} \omega_{nm} \tag{18}$$

For PVDF film the values of the piezoelectric constants can vary depending on manufacture but in general d_{33} is larger but of similar magnitude to d_{31}. This implies that the ratio q_{acc} to q_{strain} will only approximate unity for modes with a natural frequency $\omega_{nm}/2\pi$ of the order of 10^7 Hz. For the sensor to be useful for the active control of sound this is too large by a factor of at least 10^4.

The effective value of ρh_t could be increased by placing the proof mass on a number of compliant spacers, as shown in Fig. 7, which relieve base strain. The output due to strain would in this case be reduced since the effective area of the PVDF sensor would have been decreased while maintaining a similar mass layer. It is not clear whether such a strategy could increase the ratio q_{acc}/q_{strain} to be acceptable in practice.

CONCLUSION

Distributed strain sensors which measure the volume velocity of a rectangular clamped plate have been developed using PVDF film. To extend the development of volume velocity sensors to include structures that have irregular shapes, have discontinuities or exhibit whole body vibration is difficult and would in general require additional calibrated sensors. Any changes in the mechanical properties of the structure may also cause the sensor to become ineffective since the design has to be tailored to each structural application.

The development of a distributed accelerometer would overcome the major problems associated with distributed strain sensors. The shape and mechanical properties of the structure would not affect the sensor design or performance. A distributed accelerometer can be made by placing a proof mass layer over the surface of the piezoelectric but such a sensor would exhibit a sensitivity to surface strain. A dummy sensor could be used to remove this strain component but this strategy will only be effective if the strain component is reasonably small compared to the acceleration component. Unfortunately, the sensitivity to strain is likely to be many orders of magnitude larger than the sensitivity to acceleration. The sensor's sensitivity to surface strain could potentially be reduced by reducing the effective surface area of the piezoelectric material while keeping the proof mass constant. Manufacturing methods may also be able to improve the ratio of d_{33} to d_{31} and reduce the sensitivity to strain. It is, however, still unclear as to whether sufficient reductions in sensitivity to surface strain can be achieved simply and cheaply enough to made such a device workable.

REFERENCES

1. S.J. Elliott and M.E. Johnson: 'Radiation modes and the active control of sound power', *J. Acoust. Soc. Am.*, 1993, **94**(4), 2194–2204.
2. M.E. Johnson and S.J. Elliott: 'Active control of sound radiation using volume velocity cancellation', *J. Acoust. Soc. Am.*, 1995, **98**(4), 2174–2186.
3. F. Charette, C. Guigou and A. Berry: 'Development of volume velocity sensors for plates using PVDF film', *Proc. ACTIVE 95*, 1995, 241–252.
4. M.E. Johnson and S.J. Elliott: 'Experiments on the active control of sound radiation using a volume velocity sensor', *Proc. SPIE. 1995 North Am. Conf. on Smart Struct. and Mat. Vol. 2443*, 1995, 658-669.
5. S.D. Snyder, N. Tanaka and Y. Kikushima: 'The use of optimally shaped piezo-electric sensors in the active control of free field radiation, Part 1: Feedforward con trol', *J. Vib. Aco. Trans of ASME*, 1995, **3**, 311–322.
6. C.K. Lee: 'Theory of laminated piezoelectric plates for the design of distributed sensors/actuators. Part I: Governing equations and reciprocal relationships', *J. Acoust. Soc. Am.*, 1990, **87**, 1144–1158.
7. *Autochem Sensors Technical Notes*, 1987, 14.
8. J. Rex and S.J. Elliott: 'QWSIS A new sensor for structural radiation control', *Pro ceedings of the 1st International Conf. on Motion and Vibration Control*, 1992, 339–343.
9. M.E. Johnson and S.J. Elliott: 'Volume velocity sensors for active control', *Proc. Inst. Acoust.*, 1993, **15**(3), 411–420.
10. G.B. Warburton: 'The vibration of rectangular plates', *Proc. Inst. Mech. Eng.*, 1954, **168**, 371–383.

DISTRIBUTED PVDF SENSORS FOR A VIBRATING BEAM

M. SOLOOK AND S.O. OYADIJI

Dynamics and Control Research Group, The Manchester School of Engineering,
University of Manchester, Manchester M13 9PL

ABSTRACT

Distributed sensors are shaped sensors which are bonded to the surface of a structure and are continuous over the length and the breadth of the structure. They are designed on the basis of the orthogonality principle of the modes of vibration of a structure. Distributed sensors are designed to monitor specific modes of vibration of the structure and are usually made of polyvinylidene fluoride (PVDF) piezoelectric polymer films. In this paper, PVDF shaped sensors are designed for monitoring the first and the second modes of vibration of a simply supported beam. The design principle for achieving the optimal sensor shape for a particular mode of vibration is based on making the sensor width zero at locations along the length of the beam where the strain is zero. Conversely, the sensor width is maximum where the strain in the beam is maximum for the particular mode of vibration. The vibration responses of the beam as measured by the distributed sensors are compared with the measured vibration responses using an accelerometer. It is shown that by dividing a classical PVDF shaped sensor for mode 1 into two halves, the sensor can be used to monitor modes 1 and 2 of the flexural vibration of the beam.

INTRODUCTION

Considerable interest has been recently directed to active vibration control using piezoelectric sensors and actuators. The passive vibration control technique is very effective in controlling high frequency structural vibration. However, for low frequency vibration, passive control alone may not be sufficient and so it may be necessary to apply active vibration control as well. Structures incorporating active vibration control techniques can be classified in the group of smart structures.

In active structural vibration control some sensors detect the vibration of the structure while some actuators are used to provide forces or moments to control the vibration of the structure. Active vibration control methods using discrete point sensors such as accelerometers often have spill over and instability problems since these sensors detect the vibrations due to all of the modes of vibration of the structure at the same time.

This paper presents a new category of sensors called distributed sensors that are able to identify a specific mode of vibration clearly. Polyvinylidene fluoride (PVDF) is one of the common piezoelectric films which are used for distributed sensors. These films are very thin and very sensitive and they do not have much effect on the dynamics of a structure. Distributed sensors are bonded to the surface of the structure and they collect vibration data from all points of the structure simultaneously. Changing the local sensitivity and using the directional properties of a sensor film are two factors for achieving maximum total output charge generated by a distributed sensor for a desired mode of vibration and a minimum output charge for all other modes. A distributed sensor can be designed when the vibration characteristics, especially mode shapes of a structure, are fully known. However, the identification of the mode shapes of the structure usually requires the use of discrete sensors.

Lee and Moon[1] designed a sensor for a cantilevered beam. Clark, Burdisso and Fuller[2] designed two distributed sensors for a simply supported beam and a simply supported plate. Claus and Nohr Larsen[3] presented a modal sensor for a simply supported plate in which some parts of the sensor film were removed so that the sensitivity of the film varied along the X and Y directions. Gu, Clark and Fuller[4] designed and then proved experimentally a two dimensional sensor for a simply supported plate. They used two separate one dimensional sensors in the X and Y directions to monitor a two dimensional mode. Although all of these distributed sensors are different in their shapes, their designs are based on the same theory of orthogonality of the mode shapes.

Distributed sensors are more reliable and easier to use than a series of point sensors to identify a vibration mode. However, most of the distributed sensors which have been reported in the literature can only be used for monitoring one mode of vibration. In this paper, a distributed sensor which is capable of monitoring more than one mode of vibration of a beam is presented. The shaped sensor considered here is the classical semi-sinusoidal shape which is designed for monitoring the first mode of vibration of a simply supported beam. It is shown that by dividing this shaped sensor into two halves, the first mode of vibration of the beam can be monitored simply by adding the output charge from each half while the second mode of vibration of the beam is monitored by subtracting the output charge from one half from the other.

THEORY

Distributed sensors are designed on the basis of the orthogonality principle which leads to an output response for only a desired mode and no response for the other modes. The charge generated in a piezoelectric film can be determined by integrating the induced strain over the surface on which the film is bonded as:[5]

$$q = -z \int_s F(x,y) P_0 \left(e_{31} \frac{\partial^2 w}{\partial x^2} + e_{32} \frac{\partial^2 w}{\partial y^2} + 2e_{36} \frac{\partial^2 w}{\partial x \partial y} \right) dxdy \qquad (1)$$

where z is the distance of the piezo film from the neutral axis, e_{ij} are piezoelectric constants in different directions, w is displacement, s refers to surface. P_o is polarity which can be positive or negative depending on the direction of polarisation, and $F(x,y)$ is a function related to the sensor shape defined as:

$$F(x,y) = \begin{cases} 1 & \text{If point } (x,y) \text{ is on sensor electrode} \\ 0 & \text{Otherwise} \end{cases}$$

With no skew angle ($\theta = 0$ so $e_{36} = 0$), eqn (1) is simplified to the following form for a one dimensional structure such as a beam:

$$q = -z \int_0^L F(x) \cdot P_0 \cdot \left(e_{31} \frac{\partial^2 w}{\partial x^2} \right) \cdot dx \tag{2}$$

The transverse displacement of a beam can be written as the sum of its modes as:

$$w(x,t) = \sum_1^\infty A_i(t) \phi_i(x) \tag{3}$$

where $A_i(t)$ is a generalised co-ordinate, and $\phi_i(x)$ is the mode shape of the i-th mode. The strain distribution over the beam surface is proportional to the second derivative of the mode shapes $\left(\varepsilon \propto \dfrac{\partial^2 w}{\partial x^2} \right)$ and for a simply supported beam:

$$w(x,t) = \sum_1^\infty A_i(t) \sin\left(\frac{i \pi x}{L} \right) \tag{4}$$

If the shape of a distributed piezoelectric film is defined by a function that is proportional to the second derivative of a specific mode (e.g. $F(x)$ for the n-th mode) then the total output charge from the piezo sensor will be zero when the beam vibrates in all of the modes except that specific mode. This concept is shown in the following two equations mathematically.

$$q_{total} = -zb \int_0^l P_0 e_{31} \frac{\partial^2 w_l}{\partial x^2} \frac{\partial^2 w_n}{\partial x^2} dx \tag{5}$$

or

$$q = \begin{cases} -zb \int_0^l P_0 e_{31} \left(\dfrac{\partial^2 w_n}{\partial x^2} \right)^2 dx & \text{if } i = n \\ \\ 0 & \text{if } i \neq n \end{cases} \tag{6}$$

Fig. 1 A distributed sensor for the first mode of a simply supported beam.

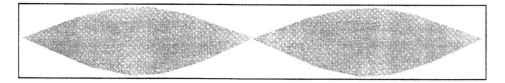

Fig. 2 A distributed sensor for the second mode of a simply supported beam.

Fig. 3 Multi-mode distributed sensor based on the mode 1 sensor shape.

Equation (6) has a non-zero value for only the n-th mode when $i=n$, which means that this sensor can monitor the n-th mode only.

Figures 1 and 2 show two modal sensors for the first and second modes of a simply supported beam which are obtained by double differentiation of the first and second mode shapes of the beam respectively. Each of these sensors monitors only the specific mode of vibration for which it has been designed. Thus, to monitor modes 1 and 2 of a simply supported beam individually will require that two shaped sensors for modes 1 and 2 be bonded to the beam. An alternative approach, which is somewhat a departure from the orthogonality principle, involves the use of the mode 1 sensor which is divided into two halves as shown in Fig. 3. In order to monitor mode 1 of the vibration of the simply supported beam, the charge outputs from each half are added to give

$$q_{total} = q_{left} + q_{right}$$

while mode 2 of vibration of the beam is given by the difference of the charge output as

$$q_{total} = q_{left} - q_{right}$$

The results show that this single sensor is very effective in detecting modes 1 and 2 of the vibration of the simply supported beam.

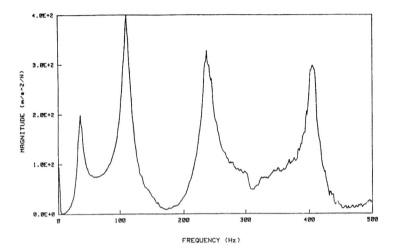

Fig. 4   Frequency response ratio of accelerometer signal to input force.

EXPERIMENTAL PROCEDURES

A rectangular piece of PVDF film was bonded to an aluminium beam of dimensions 540 x 51 x 3.2 mm thick using a double-sided tape. By means of a sharp blade, the PVDF film was cut into the mode 1 sensor shape shown in Fig. 1. The beam was simply supported on a knife-edge support and was subjected to a random vibration using a Gearing and Watson electronic shaker type GWV4/II. The input force applied to the beam by the shaker was monitored using a PCB force transducer model 208B. For comparison, the response of the beam was also monitored using an accelerometer which was located at point x = 0.28L where L is the length of the beam. The frequency response functions between the PVDF sensor output and the amplified input force excitation. and between the accelerometer output response and the input force were obtained using an Onosoki CF–350 analyser.

RESULTS AND DISCUSSION

Figure 4 shows the frequency response function obtained using the accelerometer signal. It is obvious from this figure that the beam has four natural frequencies and mode shapes within the frequency band of 0→500 Hz of the external excitation. Figure 5 shows the frequency response function obtained when the output charge from the two halves of the modified PVDF sensor are summed.

Comparing Figs 4 and 5, it is clearly seen that the PVDF sensor detects only the first mode of vibration of the simply supported beam when its two output charges are added.

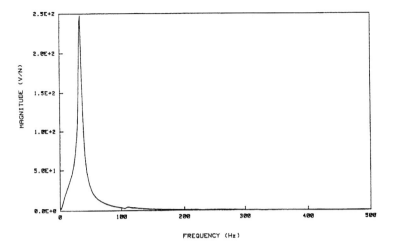

Fig. 5   Frequency response ratio of PVDF output signal to input force for mode 1 ($q_{total} = q_{left} + q_{right}$).

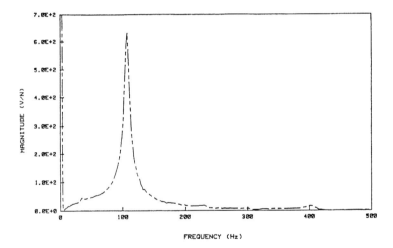

Fig. 6   Frequency response ratio of PVDF output signal to input force for mode 1 ($q_{total} = q_{left} - q_{right}$).

In effect summing the two output charges causes the modified sensor to revert back to the mode 1 sensor design and therefore validates the shaped sensor theory outlined earlier for the first mode of vibration of a simply supported beam. The frequency response function obtained from the difference of the two output charges is shown by Fig. 6. Comparing Fig. 6 with Fig. 4, it is evident that the PVDF sensor now provides the response of the second mode of vibration of the simply supported beam.

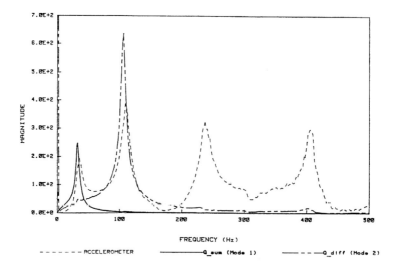

Fig. 7   Comparison of the frquency ratio of the accelerometer and
PVDF output signals to input force.

The simultaneous comparison of the three frequency response functions of the accelerometer and PVDF sensor signals is shown in Fig. 7. It is evident from this figure that the two configurations of the PVDF sensor monitor the first and second mode of vibration of the beam individually. The effective response of the other modes are quite negligible.

CONCLUSIONS

The theory for the design of distributed sensors for the spatial monitoring of selected modes of vibration of a structure, and of a simply supported beam in particular, has been outlined. The distributed sensor shapes for sensing the first and second modes of a simply supported beam have been presented. It has been shown that by dividing a mode 1 distributed sensor into two halves, the sensor can be used for selective sensing of the first and the second modes of vibration of the beam. The results show that when the sum of the output charges are used, the PVDF sensor detects the first mode only, whereas when the difference of the output charges are used, the PVDF sensor detects the second mode only.

REFERENCES

1.   C.-K. Lee and F.C. Moon: 'Modal Sensors/Actuators', *ASME Journal of Applied Mechanics,* June 1990, **57**, 434–441.

2. R.L. Clark, R.A. Burdisso and C.R. Fuller: 'Design Approaches for Shaping Polyvinylidene Fluoride Sensors in Active Structural Acoustic Control (ASAC)', *Journal at intelligent Material Systems and Structures,* July 1993, **4**, 354–365.
3. C. Larsen and P.N. Larsen: 'A New Modal Sensor for Active Vibration Control of Plates', *17th International Seminar on Modal Analysis,* KULeirven, 1992, 1717–1729.
4. Y. Gu, R.L. Clark, C.R. Fuller and A.C. Zander: 'Experiments on Active Control of Plate Vibration Using Piezoelectric Actuators and Polyvinylidene Fluoride (PVDF) Modal Sensors', *ASME J. Vibration and Acoustics*, July 1994, **116**(7), 303–309.
5. C.K. Lee and F.C. Moon: 'Laminated Piezopolymer Plates for Torsion and Bending Sensors and Actuators', *J. Acoustical Society of America*, 1989, **85**(6), 243–439.
6. R.D. Blevins: *Formulas for Natural Frequency and Mode Shape*, Van Nostrand Reinhold, NY 1979.

STRAIN MEASUREMENT WITH SURFACE ACOUSTIC WAVE (SAW) RESONATORS

A. LONSDALE AND M.J.B. SAUNDERS

Sensor Technology Ltd, P.O. Box 36, Banbury, Oxfordshire, OX15 6JB

ABSTRACT

The subject of strain measurement has previously been addressed extensively. Most of the existing transducers either rely on low voltage analogue systems (e.g. conventional resistive strain gauge), complicated mechanical assemblies by physically assessing the specimen displacement, optical systems, acoustic systems or pneumatic systems. This paper identifies a new, potentially low cost, non-contact, frequency domain strain sensor utilising SAW (surface acoustic wave) technology for surface strain measurement. The sensor has a short axial length making it flexible in terms of integration into a variety of applications, encompassing both static and dynamic strain measurement. The paper presents a technical description of the resulting strain transducer, regarding it's operation, construction and, in particular application to areas requiring strain measurement. Demonstration of the transducer performance will be addressed utilising test results from existing developed transducers.

1. BACKGROUND

The stress-strain relationship, is fundamental to the study of the mechanics of materials and is used to measure many physical quantities such as axial force, bending moment, torque, pressure, acceleration and temperature. In order to practically assess the stress state in a specimen it is imperative to measure the strain. Simple computation[1] of the strain, along with material properties, provides a method for determining the stress state. Effective practical measurement of strain has long been a problem. The following is a summary of the existing methodologies for strain assessment.

Mechanical devices evolved as the first type of strain gauge e.g. the extensometer, which relies on the displacement of levers to indicate the strain. Another example is that of the photoelectric strain sensor, which utilises the displacement of light passing through gratings separated at a set distance; sensing of variation in light intensity by photocells provides a signal indicative of the strain. On the whole, such methods are bulky, difficult to use and mostly limited to static strain analysis.

Optical techniques, such as photo-elasticity, holography or Moire methods for strain analysis prove to be both accurate and sensitive. However, the apparatus and intricacy of the optical processes generally restricts the practise of such methods to specially pre-pared laboratories.

Existing electrical devices for measuring strain can utilise any of the following meth-ods; capacitive, inductive, piezoelectric, resistive and piezo-resistive techniques can be identified as the main contributors to the solution of practical strain measurement. How-ever, all rely on the interrogation of low power density analogue signals, highly suscepti-ble to amplitude modulated noise.

Of all the techniques described, resistive strain-gauges[2] have emerged as the dominat-ing technique for strain measurement. The gauge is generally bonded onto the specimen and provides a low cost solution with the following characteristics: short gauge length, small physical size, small mass, moderate signal error resulting from temperature fluc-tuations and capability of measuring both static and dynamic strain. However translation to non-stationary applications e.g. a rotating shaft, demands expensive slip rings or large low frequency transformers.

It is apparent, therefore, that there is great demand for a low-cost, unobtrusive, high performance, non-contact strain sensor. This paper will identify a new, potentially low cost, non-contact, frequency domain strain sensor utilising SAW (surface acoustic wave) Technology[3] for surface strain measurement. The fundamental behaviour of the device will be discussed along with brief attention to it's construction. Following this, sensor performance with respect to particular application areas will be highlighted.

2. NON-CONTACT STRAIN MEASUREMENT UTILISING RAYLEIGH WAVES

2.1 SURFACE ACOUSTIC WAVES (SAW) – RAYLEIGH WAVES

In 1885, the English scientist, Lord Rayleigh, theoretically demonstrated[4] that waves could be propagated over the plane boundary between a linearly elastic half-space and a vacuum (or a sufficiently rarefied medium e.g. air), where the amplitude of the waves decays exponentially with depth. Rayleigh predicted such waves to be a major compo-nent of earthquakes, a fact to be confirmed much later in the 1920s due to the advent of seismographic recordings.

Some forty five years later, Voltmer and White, of the University of California, gener-ated such waves[5], which are more commonly referred to as surface acoustic waves (SAW) or Rayleigh Waves, on the free surface of an isotropic, elastic substrate, namely, quartz.

The action of the wave on the solid produces a pattern of displacements as illustrated in Fig. 1, where the dots indicate material particles, nominally equidistant both vertically and horizontally in the absence of wave motion. The components of displacement along the x and z axes are given by eqns (1) and (2) respectively.[6]

$$U_R = Ak_R\left[e^{-q_R z} - \left(\frac{2q_R s_R}{k_R^2 + s_R^2}\right)e^{-s_R z}\right]\sin(k_R x - \omega t) \qquad (1)$$

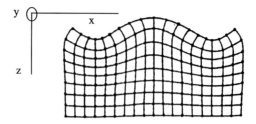

Fig. 1   Surface acoustic wave.

$$W_R = Aq_R\left[e^{-q_R z} - \left(\frac{2k_R^2}{k_R^2 + s_R^2}\right)e^{-s_R z}\right]\cos(k_R x - \omega t) \tag{2}$$

where:

A is an amplitude constant,

k_R is the Rayleigh wave number,

$k_L = \omega(\rho/(\lambda+\mu))^{1/2}$ are the wave numbers for longitudinal modes,

$k_T = \omega(\rho/\mu)^{1/2}$ are the wave numbers for transverse modes,

λ and μ are the elastic Lame constants, ρ the material density and ω the circular frequency,

$q_R = (k_R\text{-}k_L)^{1/2}$, $s_R = (k_R\text{-}k_T)^{1/2}$,

t is the time, x the in-plane displacement and z the vertical displacement.

Since the displacement components U_R and W_R in the Rayleigh wave along the x and z axes, respectively, are shifted in phase by $\pi/2$, the mode of oscillation of particles supporting a Rayleigh wave is a retrograde ellipse whose normal (vertical) displacement reaches its maximum amplitude at a depth of approximately $0.2\lambda_R$ (where λ_R is the Rayleigh wavelength) and then decays to zero within two wavelengths from the surface. It also contains a component of displacement in the plane of the solid surface.

The influence of the material properties of the surface layer of a sample on the velocity and attenuation of Rayleigh waves permits the latter to be used for the assessment of residual stresses in the surface layer, as well as the thermal[7] and mechanical properties of the surface layer of the sample. Upon this is based the application of Rayleigh waves for the non-intrusive surface testing of components[8] and of particular interest strain measurement. The requirement is for a transducer design capable of measuring such phenomena.

2.2 SURFACE ACOUSTIC WAVE TRANSDUCER CONSTRUCTION

An important property of a surface acoustic wave is that its phase velocity is considerably smaller than the phase velocity of a bulk wave in that direction. The Rayleigh wave velocity is approximately 10^5 times slower than the velocity of electromagnetic radiation in vacuo, and thus, for the same frequency, the wavelength of the elastic wave is less than

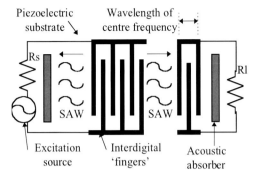

Fig. 2   Outline Of A Basic SAW Transducer

the wavelength of the corresponding electromagnetic wave by a factor of 10^5. This result has immediate importance on the geometry of the resulting gauge and is the philosophy behind surface wave technology, since the devices themselves can be much smaller than their electromagnetic counterparts, with the additional advantage that the surface wave can readily be sensed anywhere along its path. Importantly, it can be shown[4] that the Rayleigh wave does not exhibit phase velocity dispersion, i.e. there is no dependence of velocity on frequency for a homogeneous medium.

The Rayleigh waves are produced by metallic film transducers on the surface of a piezoelectric material as illustrated in Fig. 2. These transducers have an interdigital configuration which is readily fabricated using standard integrated circuit technology. The period of the fingers determines the period of the wave generated. An alternating voltage applied to the transducer causes the surface of the piezoelectric material to undulate periodically, thus producing waves. The reverse happens at the receiving transducer where the wave is converted back into an electrical signal.

The operation of a SAW transducer for strain measurement depends on the choice of a suitable piezoelectric substrate which can be attached to the material to be stressed. The stress induces a strain which can be in either a state of tension or compression. The sensitive axis of the transducer is longitudinal in the direction of wave propagation. Strain will change the spacing of the interdigital electrodes and hence the operating frequency. For example, for an excitation frequency of 500 MHz, a tensile strain of 10^{-3} will reduce the frequency by 500kHz (10^{-3} x 500 x 10^6 Hz). Conversely, a compressive strain will increase the frequency by the same amount. To function as an oscillator the element is used as amplifier feedback. The quality factor (Q) of the transducer is high, typically, 10^4, therefore, by meeting the phase and gain requirement the circuit will oscillate with very high stability, typically, one part in 10^9 is not uncommon.

From the technique described it is apparent that the output signal will be in the frequency domain as shown in Fig. 3. This has many advantages from an application viewpoint, particularly in variable speed electrical machines where low frequency signals can be easily contaminated by drive electronic generated noise. The SAW impedance, designed at the 50 W standard, means there is generally less noise than in its classical

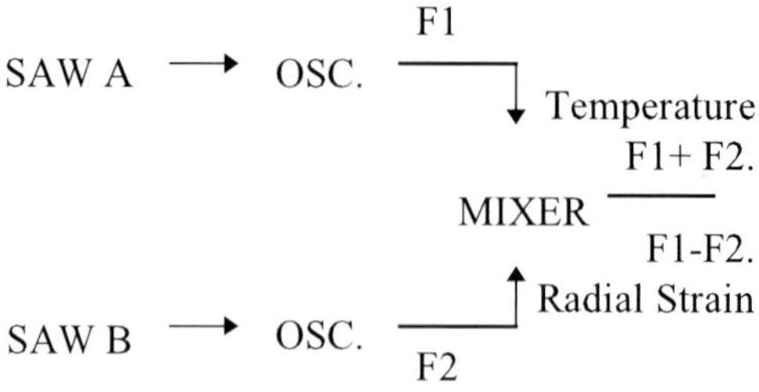

F1

SAW A → OSC. ——

↓ Temperature

F1+ F2.

MIXER ‾‾‾‾‾

F1-F2.

↑ Radial Strain

SAW B → OSC. ——

F2

Fig. 3   Schematic of frequency domain SAW signals

resistance gauge counterpart(resistance usually 200–350W). The maximum power of the SAW signal is in the region of 25mW, and so when compared with say a 10V half bridge resistive strain gauge system where the total power of the output strain signal is in the region of 1mW the SAW device offers a more robust solution to strain measurement. Piezoelectric transducers are also available for dynamic strain sensing but with the disadvantage of having a high output impedance.

2.3 APPLICATION OF SAW DEVICES FOR MEASURING STRAIN

2.3.1 Torque Measurement

Torque (radial strain) transducers are one of the more common devices used by development engineers. A knowledge of torque and rotational speed can be used to indicate power, from which the efficiency of gearboxes, transmissions, electrical machines and many other systems can readily be assessed. To apply the SAW element principle, two devices are used in a half bridge, analogous to the classic resistive strain gauging configuration; one positioned so as to be sensitive to the principal compressive strain and the other positioned to observe the principal tensile strain. Note that in the absence of bending moments and axial forces, the principal stress planes lie perpendicular to one another at 45° to the plane about which the torsional moment is applied. This is illustrated in Fig. 4.

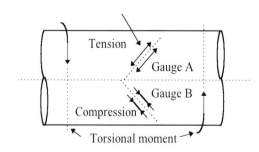

Fig. 4   SAW gauge arrangement for torque measurement

The two frequencies produced by the SAW devices are mixed together to produce the difference and/or sum signals. The difference signal is a measure of induced strain due to the twisting moment and hence, from a knowledge of the material properties and the governing equations, the torque is implied. The sum signal is a measure of shaft temperature.

Coupling of the signals to and from the machine shaft is achieved via an electromagnetic coupling device comprising two concentric hoops, separated by a suitable distance, one fixed to the shaft housing and the other fixed to the rotating shaft providing non contact interrogation which is intrinsically safe.

The primary frequency of oscillation can be chosen to lie between 100–1000 MHz with the difference frequency varying up to 1MHz. Such a transducer has the following specification:

> Resolution: 1 part in 10^6.
> Linearity: 0.1%.
> Bandwidth: >1MHz.

Figure 5 shows a typical prototype torque transducer unit as designed for application in electric power steering control. Figure 5b illustrates the mounting into a steering rack unit.

2.3.2 Force Measurement

To incorporate the SAW element principle into a force transducer, two devices are used in a half bridge, similar to the torque measurement arrangement; one positioned so as to be sensitive to the principal compressive strain and the other positioned to observe the principal tensile strain. If the force is sensed by measuring the bending strain in a bar orientated with its length perpendicular to the axis of applied force then, from classic bending beam analysis[1] the principal stress planes are located along the top and bottom surfaces of the plane of the length of the bar as illustrated in Fig. 7. The fundamental reason for incorporating a half bridge is to increase sensitivity with zero temperature compensation

Fig. 5 Prototype SAW torque transducer

Fig. 6   Typical steering column into which torque transducer is mounted

Fig. 7   Bending beam arrangement

and monitor temperature variation of the specimen material, available from the induced thermal stresses allowing correction for the variation in material properties with temperature.

One such tested force measurement apparatus shown in Fig. 8, specifically designed to measure both high level static force and low level dynamic fluctuations, exhibited good linearity in SAW device response in measuring static loads of ±100N, corresponding to approximately 200 με, as shown in Fig. 9.

The dynamic performance for the same transducer for measuring low level strain fluctuations can be seen in Fig. 10. The noise floor of the signal processing electronics is somewhere in the region of -60 dB (ignoring the effects of the induced mains harmonics due to poor screening), with a drive signal of -20 dB for 100 mN rms. So even at this low level drive input, the SNR is respectable 40 dB.

Fig. 8   SAW force transducer component

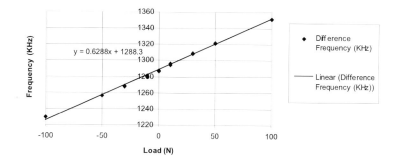

Fig. 9   Static calibration of force transducer

From static analysis it is possible to evaluate a scale factor for the strain induced per unit load. For example, in this case, the SAW gauges each shifted frequency by 40 kHz for 100 N static loading, indicating a strain of around 200 $\mu\varepsilon$ and thus a scale factor of 2 $\mu\varepsilon$ per 1 N peak loading. Combining this result with the dynamic performance it is evident that for a 40 dB 100 mN drive the equivalent strain is 0.2 $\mu\varepsilon$ rms. Therefore, the potential resolution for this application at voltage SNR 2:1, is an impressive 0.004 $\mu\varepsilon$ rms corresponding to 1 Hz in 200 MHZ sensitivity.

2.3.3 Other Applications

The fields for potential SAW strain measurement technology application are boundless. For example, pressure, level, acceleration (inc. gravity), temperature and vibration are all easily realisable, as a direct result of the demonstrated performance specification of the SAW devices.

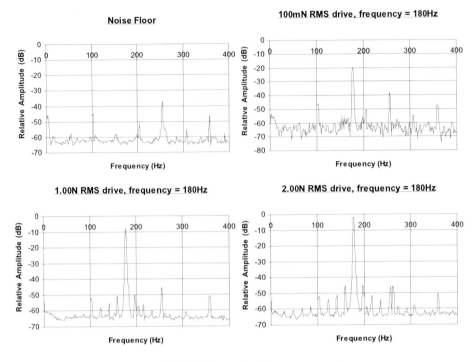

Fig. 10   Dynamic strain measurement

3. CONCLUSIONS

It has been demonstrated that SAW devices can be used as strain gauges functioning in the frequency domain. An increase in sensitivity over prior art by two orders of magnitude is not unrealistic. The low power requirements coupled with low impedance make these devices superior to conventional resistance gauges. Due to their simplicity and ease of manufacture it is not improbable that these devices will serve well into the 21[st.] Century, as the resistance gauge has served in the 20[th.] Century.

REFERENCES

1.   S. P. Timoshenko and J. N. Goodier: *Theory Of Elasticity*, Third Ed.,McGraw-Hill inc., 1970.
2.   Omega Eng. inc.: *Pressure, Strain & Force Handbook*, 1989, section-E.
3.   N. Schofield, A. Lonsdale, P. Mellor and D.I. Howe: 'An Integrated Low Cost Torque Sensor For The Direct Control Of A Brushless DC Traction Motor', *26[th.] ISATA*, Aachen, Germany, Sept., 1993, paper No.93EL082.
4.   Lord Rayleigh (J. Strutt): 'On waves propagated along the plane surface of an elastic solid', *Proc. Lon. Math. Soc.*, 1885, **17**, 4–11.

5. R.M. White and F.W. Voltmer: 'Direct piezoelectric coupling to surface elastic waves', *Applied Physics. Letters*, 1965, **7**, 314–316.
6. I.V. Viktorov: *Rayleigh and Lamb Waves - Physical theory and applications*, Plenum Press, New York, 1967.
7. D.M. Boyd and P.D. Sperline: 'Noncontact temperature measurement of hot steel bodies using an electromagnetic acoustic transducer', presented at the *Review of Progress in QNDE conference*, Williamsburg, Virginia, 21–26 June, 1987.
8. W. Sachse and N.N. Hsu: 'Ultrasonic transducers for materials testing and their characterisation', Academic Press Inc., *Phys. Acous.*, 1979, **XIV**, 277–341.

DESIGN OF 'SMART' STRUCTURES TO MINIMISE STRESSES INDUCED BY EMBEDDED FIBRE SENSORS

A.M. THORNE[†], M. HADJIPROCOPIOU*, G.T. REED*[§]
AND L. HOLLAWAY[†]

*Department of Electronic and Electrical Engineering, and [†]Department of Civil
Engineering, University of Surrey, Guildford, Surrey GU2 5XH, UK

ABSTRACT

Actuators and/or sensors embedded into a host material will disrupt the physical properties of the host. Finite element analysis was used to determine and to minimise the stress concentrations which arise in a 'smart' material system due to the embedded optical fibre sensor. An optimisation routine was used to perform a parametric study to determine the theoretical mechanical and thermal properties of the interface coating that minimise the disruption of the host material properties, due to the optical fibre inclusion. The effects of transverse tensile and thermal loading were studied, including the effect of manufacturing residual stresses. The stress concentrations in the composite host are affected by the dimensions and the mechanical and thermal properties of the interface coating. The results show that with careful selection of the interface coating properties the stress concentrations in the host material caused by the optical fibre inclusion can be reduced to levels similar to those of the pure host material. It is proposed that a set of design curves are produced for a range of host material properties so that the appropriate optical fibre coating can be selected.

1. INTRODUCTION

'*Smart*' Structures, with sensors and/or actuators embedded into or surface bonded on to a host material, are currently receiving attention worldwide. Monolithic load-bearing structures with built-in sensing systems could continuously monitor their internal strain, loading, vibration state, temperature and structural integrity. For some time it has been considered desirable to monitor strain levels internally in both composite materials and in gluelines where reinforcing materials are adhered to existing load bearing structures. The development of smart polymer materials depends upon embedding the sensory system into the host material during the fabrication process. Fibre optic sensors (FOS) have been considered to be the prime candidates for internal structure and condition monitoring of polymer composite materials.[1–3] If fibre optic sensors are incorporated into the structure the basis of an embedded sensor is formed, potentially capable of sensing strain

§ to whom correspondence should be addressed.

and temperature variations throughout its service life.

To gain confidence in the technique of embedding a sensor into a host material, and to validate the measured strain values from the sensor against a conventional electrical resistance strain gauge measurement, experimental testing of a tensile coupon specimen produced the results shown in Fig. 1. Verification of the finite element technique resulted in the strain values also shown on the plot. The detail of this work has been published by Reed *et al.*[4]

For the embedded sensor or actuator systems, the issue of the obtrusivity of the sensor to the host material arises. Obtrusivity refers to possible structural strength degradation of the host material due to stress concentrations arising from the inclusion of the sensor. Optical fibres are typically 100–300 µm in diameter and are considerably larger than the reinforcing fibres (5–10 µm in diameter) and are a significant proportion of the thickness of the gluelines it is proposed to monitor, and it is most important that the inclusion of these optical fibres does not alter the mechanical properties of the host material. The embedded optical fibres will, however, disrupt the host by causing local stress concentrations to arise.[5] The response of the material system, (host and sensor/coating), depends upon the material properties of the fibre and host material and the condition of the interface between them.[6, 7] The interface condition is important as it is the medium via which stress and strain are transferred from the host to the fibre sensor. Therefore, by careful selection of the material properties of the fibre primary coating which acts as the interface between the fibre core/cladding and the composite host, the obtrusive characteristics of the embedded fibre optic sensor can thus be minimised.[8, 9]

Most of the work carried out by other researchers has concentrated on minimising the stress concentrations caused by the stiffness mismatch between the fibre/core and the composite host. In the work presented here the effects of temperature induced stresses during the service life of the host and the manufacturing residual thermal stresses were also taken into account, and the combined effect of thermal and mechanical loading cases was also considered. Finite element (FE) modelling was employed to determine an 'optimum' theoretical coating material stiffness, thickness and coefficient of thermal expansion for an optical fibre embedded parallel to the reinforcing fibres of a unidirectional glass fibre reinforced polymer (GFRP) in order to minimise the obtrusive behaviour of the optical fibre. This 'optimum' interface material will create a system assembly (fibre/coating/host) which minimises possible degradations of the system's transverse characteristic, by reducing the stress concentrations which arise due to the fibre inclusion.

The optimisation is achieved by minimising the stress concentrations which result in the host due to stiffness mismatch between the fibre and host and also due to the residual stresses resulting from mismatches in the thermal expansion coefficients. This process is host specific and the results refer only to the host material under investigation. It is proposed that a family of 'design' curves will be produced to cover a range of possible host materials, in order to make the selection of coating parameters simpler and to avoid the need to resort to the optimisation analysis for each application.

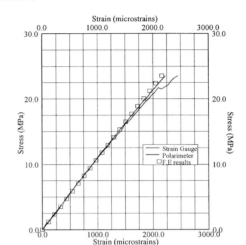

Fig. 1   ERSG Strain/optical Strain/numerical strain measurement.

1.1 Transverse Tensile Strength

A number of our current applications involve the use of unidirectional pultruded composite plates. Internal strain measurements afforded by the inclusion of the optical fibre sensor for both laboratory applications and for long term monitoring of working structures is desirable. Unidirectional laminae have very low transverse tensile strength and this does present a problem when embedding fibre optic sensors. In general, reinforcing fibres are orientated to lie parallel to the external loads, but transverse stresses cannot always be avoided and they may lead to fracture failures of the composite. Unlike the longitudinal strength and stiffness and the transverse modulus, the transverse strength is reduced in the presence of the reinforcing fibres.[10] Experimental results have also shown that the strength of composites are directly related to the micro-level composite properties, such as the presence of fibre surface treatments and fibre coatings.[11,12] Therefore the transverse tensile strength of the composite host, with the embedded sensor running parallel to the reinforcement, is very sensitive to the presence and mechanical properties of the optical fibre, and failure is thought to initiate in the host close to the interface between the sensor and the host material.[13, 14] Therefore, it is important to try and minimise the stress concentrations in the composite host, created by the optical fibre inclusion, and to maintain the transverse strength properties of the host, thereby increasing the transverse strain to failure ratio of the composite host.

1.2 Possible Application

The technology is being applied to a current investigation undertaken by the Composite Structures Research Unit of the Dept. of Civil Engineering in the University of Surrey.[15, 16] Here it is required to monitor the internal strain behaviour of a pultruded carbon-fibre reinforced composite plate bonded to a reinforced concrete beam; the concrete system is to be upgraded. In addition the internal strains in the glueline between the composite plate and the reinforced concrete beam are to be monitored. This research work employs a combination of both integrated and point sensors to monitor the laboratory beam structure. It is suggested that similar techniques could then be employed on prototype structures to continuously monitor their behaviour where the structural reliability over a number of years is required to be known. Site inspection is expensive and to detect degradation in the glueline is not currently possible as it requires expensive scanning techniques to locate possible problem areas.

2. THE FE MODEL

The investigation utilised a model of a 125 µm (diameter) silica fibre which was embedded into a unidirectional glass fibre (of 66% by weight) reinforced polymer (GFRP) plate; the optical fibre was aligned with the longitudinal direction of the glass fibre. The composite plate under evaluation is to be used as reinforcement for a concrete beam, therefore, it is important not to perturb its structural characteristics by the inclusion of the optical fibre. The model has been described in greater detail by Hadjiprocopiou *et al.*[17]

The model consists of a 2-D FE array that considers both mechanical and thermal induced stresses under the plane strain assumption. The geometry and boundary conditions of the model are shown in Fig. 2a and, since two planes of symmetry exist, only one quarter of the system was modelled. The mesh was formulated from 8-noded quadrilateral ABAQUS[18] elements (type DC2D8 for the thermal analysis and CPE8 for the structural analysis) using the pre processor of the FE package PATRAN from PDA Engineering: the mesh is shown in Fig. 2b. The detailing of the mesh enabled a range of coatings to be monitored whilst accurately modelling the interface coating material properties. The material properties used in the analysis are shown in Table 1, where E_x, E_y and E_z are the elastic modulus values in directions X, Y and Z respectively, v is the Poisson's ratio and α is the coefficient of thermal expansion. The GFRP density was calculated using the rule of mixtures[10] as shown below:

$$\rho_{GFRP} = \rho_f V_f + \rho_m V_m \tag{1}$$

where ρ refers to density, V to the fibre volume content, and m and f refer to the matrix and reinforcing fibres respectively.

The analysis of the structural system was performed using the FE package ABAQUS[18] (ver. 5.4). The edge faces of the model were restrained to remain planar and the stress concentrations, from the embedded optical fibre were computed. The parametric study undertaken involved three design variables for the coating (stiffness, thickness and thermal expansion coefficient), and the design objective was the minimisation of stress concentration in the *GFRP* host (maximum principal stress σ_{max} and maximum shear stress τ_{max}).

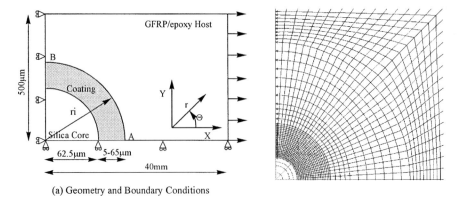

(a) Geometry and Boundary Conditions

Fig. 2   Model diagrams

3. FE ANALYSIS RESULTS

A detailed description of the FE analysis has been reported by Hadjiprocopiou *et al.*[17] The stress plots are normalised against an analysis undertaken without the optical inclusion, hence, a normalised value at any location represents the ratio of the stress state with to that without the optical fibre inclusion in the pure host. The analysis included a range of interface elastic moduli from 45 MPa, soft acrylate, through the stiffer polyimide coating, 2 GPa, to aluminium, 70 GPa and includes intermediate values which, whilst they may not yet exist, provide points to complete the picture. A uniform tensile stress was applied to the model and the resulting stress distributions were analysed.

3.1 TRANSVERSE TENSILE LOAD CASE

A plot of normalised maximum principal stress in the host against angular position from the X to Y axis, Point A to Point B, as defined in Fig. 2a, are shown in Fig. 3. The results are plotted for two of the commercially available coating thicknesses of 10 μm and 65 μm. For a coating thickness of 10 μm the modulus giving a stress closest to the norinalised ideal of 1 is that for a value of 2 GPa, but for a thicker coating of 65 μm this value increases to 10 GPa.

To gain an indication of the required coating thicknesses for the various modulus values of the coating the stress concentrations at the point A and B are plotted for various coating thicknesses. Considering the points A and B in Fig. 2a it will be seen that a tensile force applied uniformly to the model will result in a constant stress in the section and hence at both point A and B. The inclusion of the optical fibre will affect this. The 'optimum' coating thickness for a given elastic modulus is at the point where the respective curves for points A and B intersect satisfying the following criterion for selecting an 'optimum' interface material proposed by other researchers:[19, 20]

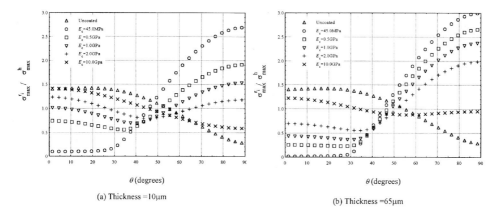

(a) Thickness =10μm

(b) Thickness =65μm

Fig. 3 Maximum principal stress concentrations (σ_{max}), distribution in the composite host at r_i, for varying interface elastic modulus (E_c).

$$\sigma_{rr}^{host}(r_i, 0°) = \sigma_{\theta\theta}^{host}(r_i, 90°) \tag{2}$$

Plots of normalised maximum normal stress in the host against angular position from the X to Y axis, (Point A to Point B), as defined in Fig. 2a, are shown in Fig. 4. In Fig. 4a a coating thickness of 7 μm is optimum for a coating modulus of 1 GPa and a coating thickness of ~12 μm for the 2 GPa polyimide coating. The coating modulus of 10 GPa would require a thickness in excess of 70 μm to achieve the desired optimum. Such a coating thickness would result in a fibre diameter which is a significant proportion of a plate 1 mm thick and could well be undesirable for this application. All the findings are confined by looking at the stress compatibilities in the YY direction plotted in Fig. 4b.

Considering out-of-plane failures in order to confirm the above, Fig. 5 shows the plot of maximum principal stress against the angular position for a number of coating thicknesses for two coating modulus values. The 2 GPa, polyimide, coating modulus suggests a coating thickness of between 10 and 20 μm, this value yielding stress being closest to the desired ratio of 1, whilst the stiffer 10 GPa coating modulus suggests a coating thickness in excess of 65 μm, in order for this ratio to tend to unity. This supports the above findings from Fig. 4.

Figure 6 shows the plot of maximum shear stress against the angular position for a number of coating thicknesses for two coating modulus values. If the dominant stress condition is shear, as undergone by a specimen in torsion, it is possible that the effect of the inclusion may be different. In Fig. 6 the maximum shear stress is plotted against angular position for the two coating thicknesses. It is evident that the coating thickness of 65 μm with a coating modulus of 10 GPa matches best the optimised condition of 1. This is not consistent with the findings for the principal stress condition and would indicate a need to consider the stress regime to be encountered and to tailor the coating properties accordingly.

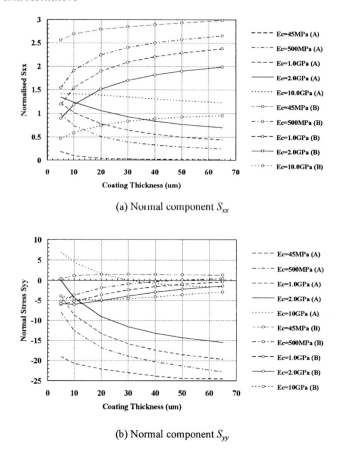

(a) Normal component S_{xx}

(b) Normal component S_{yy}

Fig. 4   Stress concentrations in the composite host at point A and B as a function of the coating thickness (T_c) and elastic modulus (E_c).

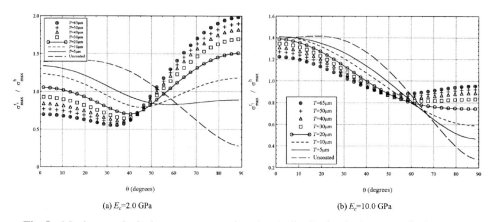

(a) E_c=2.0 GPa

(b) E_c=10.0 GPa

Fig. 5   Maximum principal stress concentrations (σ_{max}), distribution in the composite host ar r_i, for varying interface thickness (T_c).

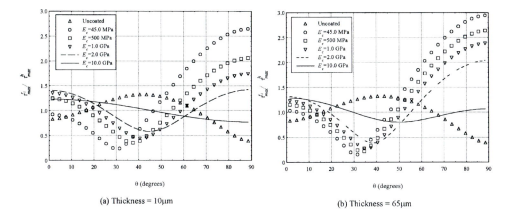

(a) Thickness = 10μm

(b) Thickness = 65μm

Fig. 6 Maximum shear stress (τ_{max}) concentration distribution in the composite host at r_i for varying interface elastic modulus (E_c).

3.2 THERMAL LOAD CASE

The material properties used for the thermal analysis can be seen in Table 1. The first analysis undertaken neglected the residual stress concentrations developed in the material as a result of the manufacturing process, but these have been considered in the next section. The thermal analysis performed was primarily concerned with the service life of the *GFRP* plate and how it is affected by the temperature changes during a working day. The stress concentrations generated in the composite host, due to the embedded optical fibre, were studied for the case when the temperature changes from -20 °C to +30 °C (one day cycle, a total change of 50 °C).

Stress concentrations will arise in the composite host due to the mismatch of the coefficients of thermal expansion between the material system components (core/cladding, interface coating and *GFRP*). For the pure host material, providing there are no external restraints, expansion due to a change in temperature will result in zero radial or hoop stresses in the section. However, due to the mismatch of coefficient of thermal expansion of materials, both a hoop stress and a radial stress will be propagated by the inclusion. Optimisation of the coating properties will result in this increase being as close to zero as possible at the coating/host interface. The thermal coefficient of expansion of the interface coating was varied (from $10\times10^{-6}/°C$ to $400\times10^{-6}/°C$) and two interface thicknesses were studied (10 μm and 65 μm). Figure 7a and b show the variation in the thermally induced radial stress along the x-axis away from the fibre centre and Figs 7c and d show the hoop stress concentrations. It can be seen that for $T_c = 10$ μm the stress components decrease (in the core/cladding region as well as the *GFRP*) as the coefficient of thermal expansion (α) is increased, whereas for $T_c = 65$ μm, as α is increased, the stress concentrations in all regions increase. However, it is evident that there is an optimum value of α, as in both cases the variation goes from tension to compression, and it will be

Table 1   Material properties used in the analysis

Properties	Core/cladding	Interface coating	GFRP host
E_x, E_y (GPa)	72.9	0.045 to 72.9	10.0
E_z (GPa)	72.9	0.045 to 72.9	48.9
υ	0.17	0.34	0.288
α_x, α_y (°C^{-1} x 10^{-6})	0.45	10-100	13.38
α_z (°C^{-1} x 10^{-6})	0.45	-	8.60
Conductivity (W m^{-1} °C^{-1})	1.02	0.20	0.30
Density (10^3 Kg m^{-3})	2.56	1.11	2.139

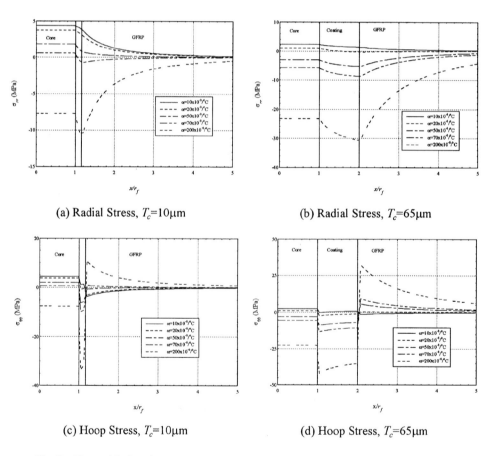

(a) Radial Stress, T_c=10μm

(b) Radial Stress, T_c=65μm

(c) Hoop Stress, T_c=10μm

(d) Hoop Stress, T_c=65μm

Fig. 7   Thermal induced stress concentrations due to a 50 °C temperature change for the polyimide coating (Ec = 2.0 GPa) for varying thermal expansion coefficients.

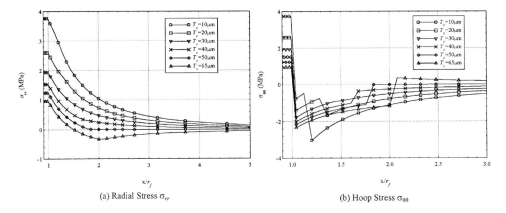

(a) Radial Stress σ_{rr} (b) Hoop Stress $\sigma_{\theta\theta}$

Fig. 8 Thermal induced stress concentrations due to a 50 °C temperature change for the polyimide coating (E_c = 2.0 GPa, α = 20x10⁻⁶ °C) for varying interface thickness (T_c).

at the zero value that the material with the inclusion behaves as the pure material. The thermal induced stresses in the core/cladding region of the fibre are also of particular interest as for some strain sensing techniques it is desirable to avoid thermally induced strains or in the case of cure monitoring, high thermally induced strains are desirable in order to achieve high sensitivity. Polyimide coatings are of particular interest and in Fig. 8 the thermal stress concentrations for a polyimide coated fibre for varying interface thickness are shown (E_c =2.0 GPa and α = 20x10⁻⁶/°C). It can be seen that for T_c = 50 µm the hoop and radial thermal stress concentrations in the *GFRP* host are rninimised (almost zero), and the core/cladding stress concentrations are greatly reduced.

Following the conclusions made above a detailed thermal analysis was also performed to determine the variation of the radial and hoop stress concentrations in the *GFRP* host at point A (see Fig. 2a) and the results are shown in Fig. 9. The graphs obtained are for a coating thickness of 12 µm which was one of the 'optimum' geometries suggested in the previous section. It can be seen that, as the interface coating elastic modulus is increased the thermally induced stress concentrations increase. The stress concentrations also increase when the thermal expansion coefficient is increased. This is in opposition to the transverse tensile analysis results where the low modulus coating had the worst effect on the *GFRP* host. Thus if the objective is only to minimise residual thermal stresses this may be achieved by making the interface coating as compliant as possible. Such a compliant coating)however, will severely degrade the transverse performance of the host, as shown in section 3.1, and will need to be considered.

However, one significant observation is that for an interface coating thermal expansion coefficient, α~65x10⁻⁶/°C ($\pm$2.5%), the stress concentrations in the composite host are a minimum (almost zero) for all values of elastic modulus. This is significant as thermal stress concentrations are now almost eliminated but a very wide spectrum of coating stiffness values can be investigated for different optimisation criteria. This result is only for the host material considered and will change if the coating thickness considered also changes.

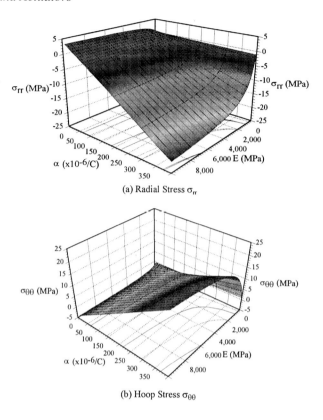

(a) Radial Stress σ_{rr}

(b) Hoop Stress $\sigma_{\theta\theta}$

Fig. 9   Variation of thermal induced stress concentrations at $0°$ (point A) due to a 50 °C temperature change as a function of the coating elastic modulus and coefficient of thermal expansion.

3.3 EFFECT OF MANUFACTURING INDUCED THERMAL STRESSES

Thermal residual stresses are generated during the processing of glass fibre/epoxy composites, particularly during cooling from the cure temperature. These stresses are the result of the different mechanical and thermal expansion characteristics of the optical fibre and the surrounding epoxy/fibre composite host. The residual stresses are important in as much as they affect the overall mechanical behaviour of the composites. If these stress concentrations are significant, failure may be initiated at external loads which are lower than predicted. In this section the effect of these residual thermal stresses is considered.

The material system was assumed to cool down from a curing temperature of 160 °C to 20 °C with a state of zero stress and strain at the start of the cooling process, simulating the conditions during the manufacturing of the pultruded *GFRP* plates. The model was then subjected to a transverse tensile load, because, as mentioned previously, the objective is to minimise all stress counteractions not just thermal stresses. The glass transition temperature is assumed to occur just below the curing temperature so that

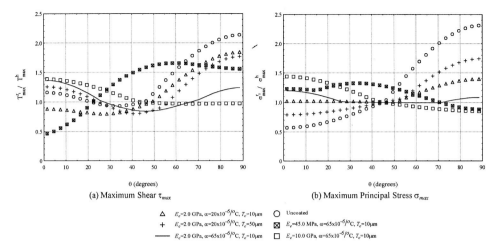

Fig. 10 Normalised stress concentration variation in the GFRP host subjected to a transvere tensile load taking into consideration manufacturing induced thermal stress due to cooling down from 160 °C to 20 °C.

elastic conditions apply over the whole time that cooling is occurring. The conclusions made from the transverse tensile and thermal loading cases were combined and the material systems studied were the ones termed as 'optimum'.

Figure 10 shows the normalised maximum principal and maximum shear stress concentrations in the GFRP host against angular position. The elastic modulus values of 45.0 MPa, 2.0 GPa and 10.0 GPa were used for the interface; the uncoated case was also considered. The thickness used was $T_c = 10$ μm and the coefficient of expansion was set to $\alpha = 65 \times 10^{-6}/°C$; this is the 'optimum' value for the 10 μm coating thickness found in the thermal analysis. The analysis also included the coefficient of expansion for the polyimide coating ($\alpha = 20 \times 10^{-6}/°C$) for two thickness values ($T_c = 10$ and 50 μm).

It can be seen that the soft acrylate coating ($E_c = 45$ MPa) causes the highest stress concentrations in the host, although they are reduced when compared to the result shown in Fig. 3; this result shows the effect of the thermal stresses. For the polyimide coating the thermal analysis suggested an optimum thickness of 50 μm for the actual thermal expansion coefficient of $\alpha = 20 \times 10^{-6}/°C$ of the material. However, it can be clearly seen that increased interface thickness, in order to reduce thermal stresses, is not justified as the stress concentrations increase when compared to the 10 μm coating thickness. Hence coatings of smaller thicknesses are favoured for embedded FOS applications. The stress concentrations are also reduced for the 10.0 GPa coating when compared to the results of Fig. 3 and Fig. 6. However, for the combined effect of residual thermal stresses and transverse load the stress concentrations in the host (σ_{max}, τ_{max}) are minimised by the proposed optimised coating, $E_c = 2.0$ GPa, $T_c = 10$ μm, $\alpha = 65 \times 10^{-6}/°C$. Compared to the compliant coating the stress concentrations are reduced by 1.5 times, they also satisfy the condition given in eqn 2.

4. CONCLUSIONS

The results obtained from the FE analysis indicate that an 'optimum' interface coating exists such that the combined properties of the optical fibre coating and *GFRP* host minimise the stress concentrations in the material system. This 'optimum' interface results in a system whose mechanical properties are similar to that of the pure *GFRP* host and this was shown to be the case with the minimisation of the maximum principal, stress and maximum shear stress concentrations in the *GFRP* host.

If a maximum principal stress failure criterion is used, the results suggest that the transverse strain to failure of the *GFRP* may be increased by 1.5 times by the use of the optimised coating.

The transverse loading case results indicate that an interface coating with an elastic modulus value of $E_c = 2.0$ GPa and a thickness of $T_c \sim 12$ μm will minimise stress concentrations in the host.

In the case of maximum shear stress concentrations the stiffer coating of $E_c = 10.0$ GPa gave the best results and the stress variations in the *GFRP* host around the coating were decreasing as the coating thickness was increased. The parametric study for a transverse tensile load showed that compromises need to be made according to the design requirements and the possible service life of the material system.

Initially the thermal analysis was performed assuming that no residual stresses develop in the material system during the cool down process after manufacturing. The thermal loading results indicate that stress concentrations in the *GFRP* host are affected by the coefficient of thermal expansion of the coating as well as its stiffness. Furthermore, by varying the thermal expansion coefficient, the residual stress concentrations in the *GFRP* host as well as the corel cladding region of the optical fibre can be minimised.

A detailed analysis for a 12 μm interface coating thickness suggests that, for a coating expansion coefficient of $\alpha = 65 \times 10^{-6}/°C$, the stress concentrations in the *GFRP* host are minimum for all values of coating elastic modulus. This is significant as thermal stress concentrations are now almost eliminated but a very wide spectrum of coating stiffness values can be utilised for different optimisation criteria, either reducing the fibre sensor obtrusivity or enhanc~the sensor performance. However the result is specific for the material system considered in this study.

When the effects of residual thermal stresses, as a result of manufacture, are combined with those for a transverse tensile load the use of the 'optimised' coating parameters established from the two loading cases further minimised the stress concentrations in the *GFRP* host.

It was observed that optimising for load dominated optimising for thermal stresses.

The study suggests that a fibre optic sensor can be designed not only for a specific host but also for a specific sensing application.

REFERENCES

1. R. Davidson and S.S.J. Roberts: 'Internal monitoring of structural composite materials through optical fibre sensors', *Measurment,* 1993, **11**, 347–360.
2. R.M. Measures: 'Smart composite structures with embedded sensors', *Composites Engineering*, 1992, **2**(5–7), 597–618.
3. M. Hadjiprocopiou, G.T. Reed, L. Hollaway and A.M. Thorne: 'Experimental results and finite element modelling of an embedded polarimetric sensor', *Composites*, 1995, **26**(11), 775–778.
4. G.T. Reed, M. Hadjiprocopiou, A. Moller, H. Garden, A.M. Thorne and L. Hollaway: 'Low Cost Dynamic Optical Strain Measurements in Glass Reinforced Polymer *Advanced Composite Letters*, 1994, **3**(3), 83–88.
5. R. Davidson and S.S.J. Roberts: 'Finite element analysis of composite laminates containing transversely embedded optical fibre sensors', *First European Conference on Smart Materials and Structures,* IOP, 1992, 115–122.
6. R. Davidson, D.H. Bowen and S.S.J. Roberts: 'Composite materials monitoring through embedded fiber optics', *International Journal of Optoelectronics*, 1990, **5**(5), 397–404.
7. S.S.J. Roberts and R. Davidson: 'Short term fatigue behavior of composite materials containing embedded fiber optical sensors and actuators', *First European Conference on Smart Materials and Structures*, IOP, 1993, 255–262.
8. G. Carman and J.S. Reifsnider: 'Analytical optimization of coating properties for actuators and sensors', *Journal of Intelligent Material Systems and Structures*, 1993, **4**, 89–97.
9. A. Dasgupta and J.S. Sirkis: 'Importance of coatings to optical fibre sensors embedded in "smart" Structures', *AIAA Journal*, 1992, **30**(5), 1337–1343.
10. L. Hollaway: 'Polymer Composites for Civil and Structural Engineering', *Blackie Acadenuic & Professional Glasgow,* 1993.
11. M.S. Madhukar and Drzal: 'Fiber-Matrix Adhesion and its Effect on Composite Mechanical Properties: I. Inplane and Interlaminar Shear Behavior of Graphite/Epoxy Composites', *Journal of Composite Materials*, 1991, **25**, 932–957.
12. M.S. Madhukar and Drzal: 'Fiber-Matrix Adhesion and its Effect on Composite Mechanical Properties: II.Longitudinal (0°) and Transverse (90°) Tensile and Flexure Behavior of Graphite/Epoxy Composites', *Journal of Composite Materials*, 1991, **25**, 958–991.
13. S.S.J. Roberts and R. Davidson: 'Mechanical properties of composite materials con taining embedded fibre optic sensors', *SPIE*, 1991, **1588**, 326–241.
14. D.W. Jensen, J. Pascual and J.A. August: 'Performance of graphite/bismaleimide laminates with embedded optical fibers. Part 1: uniaxial tension', *Smart Materials & Structures,* 1992, **1**(1), 24–30.
15. H.N. Garden, L.C. Hollaway and A.M. Thome: 'A preliminary evaluation of carbon fibre reinforced ploymer plates for strengthening reinforced concrete', *Proceedings Institution of Civil Engineers, Structures and Bridges,* Vol 122, Issue 2, May 1997.

16. R.I. Quantrill, L.C. Hollaway and A.M. Thorne: 'Predictions of the maximum plate end stresses of FRP strengthened beams', *Magazine of Concrete Research*, 1996, **48**(177), 331–351.

17. M. Hadjiprocopiou, G.T. Reed, L. Hollaway and A.M. Thorne: 'Optimization of fibre coating properties for fiber optic smart structures', *Smart Mater. Struct.*, 1996, **5**, 441–448.

18. ABAQUS. 1994 Theory and user's manual, Version 5.4 Hibbit, Karlsson and Sorensen Inc., Providence, Rhode Island, USA.

19. S.W. Case and G.P. Carman: 'Optimization of Fiber Coatings for Transverse Perform ance: An Experimental Study', *J. Composite Materials*, 1994, **28**(25), 1452–1466.

20. L.D. Tryson and J.L. Kardos: 'The Use of Ductile Innerlayers in Glass Fiber Rein- forced Epoxies', *36th Annual Conference, Reinforced Plastics/Composites Institute The Society of the Plastics Industry Inc.*, 1981, 1–5.

SPUTTER-DEPOSITED ZINC OXIDE THIN FILMS FOR SAW STRAIN SENSORS

V.P. KUTEPOVA, D.A. HALL, F.R. SALE AND S.P. SPEAKMAN*

Materials Science Centre, University of Manchester, Grosvenor Street, Manchester M1 7HS, UK
*Thin Film Technology (Consultancy) Ltd., 7 Chapel Drive, Little Waltham, Chelmsford, Essex CM3 3LW

ABSTRACT

X-ray diffraction (XRD) and scanning electron microscope (SEM) examination have been used to investigate the microstructural characteristics of zinc oxide thin films. The films were prepared by reactive RF magnetron sputtering on Si/SiO_2, $Si/SiO_2/Al$ and stainless steel substrates. The characteristics of the ZnO film structures are discussed in relation to the deposition parameters, which should lead to effective utilisation of this material for acoustoelectronic device applications. A surface acoustic wave (SAW) resonator structure with arrays of reflecting metal strips is proposed for use as a strain sensor.

1. INTRODUCTION

SAW sensors show great promise for a variety of different applications due to a combination of high Q (giving high sensitivity), high operating frequency (enabling novel, non-contact applications), and the quasi-digital output signal in the form of a frequency shift. As a result of these desirable characteristics, there is a need to develop low loss, temperature stable piezoelectric materials which can serve as the means for the excitation and detection of surface acoustic waves in various SAW devices (e.g. resonators, delay lines). This need is commonly met by temperature stable ST-cut quartz single crystals in many device applications. In contrast, layered structures of piezoelectric thin films deposited onto various substrate materials offer greater flexibility in terms of functional properties and, in many cases, a more practical means of implementing a SAW sensor. For films with a thickness smaller than one wavelength, the SAW velocity and coupling coefficient exhibit significant variations with thickness and depend on the physical properties of both the thin film and the substrate, presenting a wide variety of design choices.

Among the various piezoelectric films commonly used, zinc oxide deposited on a silicon substrate is often identified as being the most attractive for sensor applications. ZnO has a relatively high coupling coefficient and can be sputtered in an oriented polycrystalline form with structural quality close to that of a single crystal.[1-5] Silicon is commonly chosen as being the most readily available substrate material and due to the potential for the development of integrated SAW components. Furthermore, combining ZnO and Si with an intermediate silicon oxide layer in a SAW device can lead to a near zero temperature coefficient of the resonant frequency.

In a SAW device, the film properties are influenced by changes in the environment, leading to a variation in the SAW velocity, wavelength or attenuation. This effect can be quantified by measuring the frequency shift of the SAW device for monitoring stress/ strain, pressure, gases/vapour, acceleration and temperature. For a SAW strain sensor, the relative or fractional frequency change depends on the change in velocity and elongation of the propagation path of the SAW device due to changes in strain. The resulting frequency shift for bulk materials such as crystalline quartz or lithium niobate is only slightly different from the value obtained when only elongation of the propagation path due to strain is considered. In contrast, it has been reported that the effect due to the velocity variation in ZnO thin films far exceeds that due to elongation.[6] Therefore, it is expected that thin film sensors will have higher sensitivity to strain than equivalent devices based on bulk piezoelectric materials and that the sensitivity will be different for different substrate materials.

In this paper we report the results of a systematic study concerned with the deposition of ZnO thin films by reactive RF magnetron sputtering. XRD and SEM were employed to characterise the variations in film structure as a function of the deposition parameters, in order to determine the optimum film processing parameters for acoustoelectronic device applications. The quality of ZnO thin films used for SAW devices should be close to that of single crystals; they should have a smooth surface and comprise high density, well oriented crystallites. These factors are determined by: (a) the sputtering technique employed; (b) key process parameters; (c) the substrate material; and (d) the substrate surface characteristics.

2. EXPERIMENTAL PROCEDURES

2.1 MAGNETRON SPUTTERING SYSTEM

The zinc oxide thin films were deposited by reactive sputtering, using an RF magnetron sputtering system based on an Edwards Coating System E306A, onto Si, Si/SiO$_2$, Si/ SiO$_2$/Al and stainless steel substrates. The planar magnetron, measuring 50 mm in diameter, was cooled by running water. The target was metallic zinc (99.95 purity, 50 x 50 x 1.5 mm^3, Goodfellow Company), bonded to the electrode with conductive silver-loaded epoxy cement. The vacuum chamber employed was a Pyrex bell-jar type, 300 mm in diameter and 370 mm in height, which was evacuated by a cryo pump–rotary pump combination to a pressure of 7 x 10^{-7} mbar. The pressure in the chamber was measured using a Pirani–Penning gauge combination.

High-purity argon and oxygen were used as the sputtering and reactive gases respectively, the oxygen being introduced close to the surface of the growing film. Both gases were controlled using a mass flow controller, DynaMass KM-4, which operated two separate controlled leak valves in order that the partial pressure of each gas could be adjusted independently. Firstly, oxygen was introduced into the chamber and its required partial pressure set. After that, argon was introduced until the required pressure was reached. The sputtering pressure in these experiments was varied between 1.8×10^{-3} and 9.3×10^{-3} mbar. The lowest value employed was limited by the ability to maintain a stable plasma in the sputtering system below this pressure.

Substrate heating was provided by a low voltage dichroic lamp directed towards the base of the substrate holder, providing a maximum substrate temperature of around 400 °C. However, the residual pressure in the chamber was found to increase dramatically when the temperature exceeded 300 °C due to outgassing from the substrate holder framework, the target clamping arrangement and the other parts of the chamber. Therefore the substrate temperature in these experiments did not exceed 300 °C and the residual pressure prior to deposition was generally in the region of 2×10^{-6} mbar. A temperature gradient was found across the area of the substrate (with a typical size of 22 x 25 mm), the region with uniform temperature being the central part of the substrate with a diameter near 12 mm. For good thermal contact, the substrates were mounted directly onto the surface of the heated block.

X-ray diffraction measurements were carried out using a Philips X'PERT diffractometer, using Cu Kα radiation, in order to determine the phase composition, crystal structure, and the crystallographic orientation of the deposited films. The surface topology and the cross-sectional structure of the films were examined using a Philips SEM 505.

2.2 PROPOSED DESIGN OF SAW STRAIN SENSOR.

From previous investigations[7,8] it was found that the highest value of coupling coefficient, is dependent on the thickness of ZnO and SiO$_2$ layers and crystallographic orientation of the Si substrate, and for thin ZnO films can be achieved by placing the conductive IDTs (interdigitated transducers) on the top of the ZnO film and providing a ground plane between the ZnO and SiO$_2$ layers. Figure 1 shows a schematic diagram of such a SAW one-port resonator structure (typically employing aluminium electrodes for both the IDT and the ground plane) on a piezoelectric ZnO layer deposited on a thermally oxidised Si substrate. Electrode patterns of this type are readily fabricated using 'lift-off' photolithographic method. Here, the ZnO and SiO$_2$ layer thicknesses, normalised to wavenumber k, $(k = 2\pi/\lambda)$, are denoted by h_1k and h_2k respectively.

In fabricating the devices, the (100) cut of silicon was used as the substrate, with wave propagation being along the <010> direction. This pure mode direction results in the decoupling of the transverse mode from the Rayleigh-like mode, permitting the selective excitation of the latter by means of an interdigital transducer. The purpose of the SiO$_2$ layer is to provide temperature compensation, since the temperature coefficient of resonant frequency is known to vary as a function of the thickness of the intermediate SiO$_2$ layer.[8,9] When temperature compensation is desired, the SiO$_2$ layer thickness is fixed relative to the wavelength with $h_2k=0.47$.

The aluminium shorting plane of 0.1 mm thickness was evaporated under vacuum

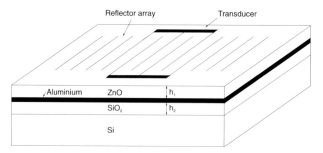

Fig. 1   Proposed layered structure of SAW thin film strain sensor.

onto the SiO$_2$ layer. This conductive layer enhances electromechanical coupling by shielding mobile charge carriers in the silicon substrate, which prevents them from interacting with the electric field originating in the ZnO layer.[10] The wave-carrier interaction can cause a degree of Q degradation due to acoustoelectronic attenuation.

Since a maximum value of the coupling coefficient K_{SAW} is not necessary in resonators, it is advantageous to use the minimum thickness of ZnO which will yield acceptable values of coupling. The use of relatively thick SiO$_2$ layers, motivated by temperature compensation considerations, has been found to yield acceptable values of K_{SAW} with thin layers of ZnO.[9] In Table 1, typical thicknesses of the temperature stabilising SiO$_2$ layer, the ZnO film and the metal electrode strips are given for two different SAW wavelengths (corresponding to two different resonant frequencies). It is evident that for a coupling coefficient near to 0.3–0.4 %, the required ZnO film thickness is typically 0.8–1.6 mm for devices operating at 100 MHz, and 0.5–1.0 mm for devices operating at 200 MHz.

Several advantages result from the use of relatively thin ZnO layers. Firstly, it is easier to grow thin films of good quality than thicker ones; surface roughness is found to increase with ZnO film thickness. Secondly, the propagation loss is diminished by using a thinner ZnO layer. Furthermore, for thinner films the SAW velocity is higher, which leads to the production of higher frequency devices at a given wavelength.

Table 1   Typical thicknesses of the SiO$_2$ layer, the ZnO film and the metal electrode strips for SAW wavelengths of 40 and 24 μm

$K_{SAW}{}^2$	kh (ZnO)	h (ZnO) μm	kh (SiO$_2$)	h (SiO$_2$) μm	kh (strip)	kh (strip) μm	V_{SAW} ms^{-1}	f_0 MHz
			Wavelength $\lambda = 40$ μm					
0.008	0.25	1.6	0.47	0.3	0.04	0.25	4200	105
0.006	0.12	0.8					4410	110
			Wavelength $\lambda = 40$ μm					
0.008	0.25	1.0	0.47	0.3	0.04	0.15	4200	175
0.006	0.12	0.5					4410	184

3. RESULTS

The ZnO films were sputtered under a variety of different deposition conditions in an attempt to establish the optimum conditions for the deposition of high density, fine grained *c*-axis oriented films onto the various substrates under investigation. Most of the deposited films exhibited a very smooth surface and good adherence to the substrates. The deposition conditions of some of the ZnO films produced during the experiments are given in Table 2. The target-substrate distance was fixed at a value of 6 cm for each sputtering process.

Table 2   Deposition conditions of selected ZnO films

Sample identifier	Sputtering pressure 10^{-3} mbar	Oxygen partial pressure 10^{-4} mbar	Substrate temperature ° C	RF power W	Deposition time min	Film thickness µm
1 (Si)	9.3	5.0	300	90	15	0.6
2 (Si/SiO₂)	9.3	5.0	300	90	31	1.2
3 (Si/SiO₂)	9.3	6.8	300	90	50	1.0
4 (Si/SiO₂/Al)	9.3	6.8	300	90	31	1.0
5 (St. steel)	9.3	8.0	300	90	35	1.0
6 (Si/SiO₂/Al)	9.3	5.0	300	70	30	1.0
7 (Si)	7.4	9.0	300	70	15	0.35
8 (Si)	7.4	9.0	300	110	15	0.5
9 (Si/SiO₂)	7.4	9.0	300	140	35	1.5
10 (Si)	2.4	3.6	250	50	15	0.45
11 (Si)	2.4	4.2	250	50	20	0.45
12 (Si)	1.8	3.6	300	50	15	0.5
13 (Si)	1.8	5.0	200	90	15	0.4
14 (Si)	1.8	5.0	250	90	40	1.2
15 (Si)	1.8	5.0	300	90	23	0.6
16 (Si)	1.8	3.6	250	90	30	1.2
17 (Si/SiO₂)	1.8	7.4	250	90	50	1.0

3.1 X-Ray Diffraction

Figure 2 shows XRD spectra for the different parts of Film N1, in the central and edge regions, where the temperatures were around 300 and 250 °C respectively. This film was grown at the highest Argon rate flow in our experiment, equal to 120 sccm, corresponding to a partial pressure of 8.8 x 10^{-3} mbar. The film grew with a preferential orientation of the *c*-axis perpendicular to the surface. The position of the (002) peak for the central part of the film was closer to that of ZnO powder than was found for the edge region. The central part of the film showed a higher crystalline content, with the intensity of the (002) peak for the central region being approximately 1.5 times higher than that of the edge. With the increase of deposition time (Fig. 2, Film N2), the texture became more *c*-axis oriented and the (002) peak was sharper, which can be an indication of a larger crystallite size or smaller strain in the lattice.

XRD spectra of the central and edge regions of films grown at the smallest argon flow rate of 12 sccm, corresponding to a partial pressure of 1.5 x 10^{-3} mbar, are shown in Fig. 3. In all other respects, the deposition parameters were identical to those of the films discussed previously. It appears that the *c*-axis orientation of these films was favoured by the lower temperature of 250 °C (the edge part of the Film N15). At a lower substrate temperature of 200 °C (Film N13, Fig. 3), the degradation of the XRD spectrum is obvious. The film structure did not improve significantly for longer deposition times (Film N14, Fig. 3).

XRD line broadening and a shift of the (002) peak to a position below that reported for pure ZnO powder were evident for all of the films discussed above, suggesting a certain

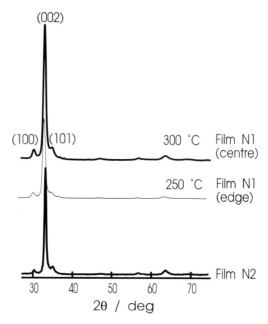

Fig. 2   XRD patterns of ZnO films prepared at a high sputtering pressure of 9.3 x 10^{-3} mbar.

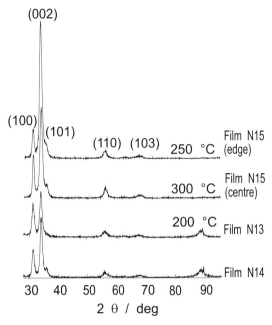

Fig. 3 XRD patterns of ZnO films prepared at a low sputtering pressure of 1.8×10^{-3} mbar.

degree of strain in the crystal lattice. Therefore, the oxygen partial pressure and RF power were varied in order to determine the optimum processing parameters for the deposition of a strain free, well oriented structure. It is evident from Fig. 4 that when the ratio of oxygen partial pressure to RF power increased (Films N4 and N6) the XRD spectra consisted of only one very sharp peak, reflected from the (002) plane. Its position was very close to that of a stress-free single crystal of ZnO. The calculated values of the lattice constant c for Films N4 and N6 were 0.51195 and 0.52084 nm respectively. The lattice constants for single crystal ZnO have been reported as c=0.52069 ±0.00442 nm and a=0.32495 ±0.00442 nm.[11] This indicates the non-strained lattice in films N4 and N6, with the c value being smaller than that for the films discussed above.

Films produced using processing parameters close to these were found to exhibit reproducible XRD spectra, independent of the substrate nature. As a result, it was possible to deposit c-axis oriented ZnO films on stainless steel by increasing the oxygen partial pressure slightly to 8×10^{-4} mbar (Film N5, Fig. 4). However, some degradation of the film orientation was evident for Film N3 which was attributed to contamination; for this deposition experiment it was observed that the residual pressure in the chamber increased from a typical value of 2×10^{-6} mbar for the other films to 3.5×10^{-6} mbar. The films N4, N5 and N6, grown onto Si/SiO$_2$/Al and stainless steel substrates, had thicknesses of about 1 μm, and were considered to be suitable for further investigations of their piezoelectric properties and for the fabrication of the chosen SAW resonator structure.

Several films were fabricated at a reduced sputtering pressure of 7.4×10^{-3} mbar and a rather high oxygen partial pressure of 9×10^{-4} mbar. It is clearly evident from the XRD

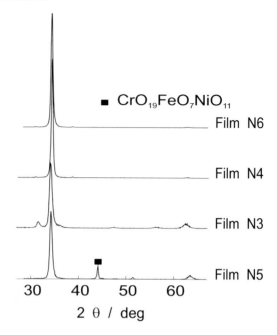

Fig. 4   XRD patterns of ZnO films prepared at a high sputtering pressure of 9.3 x 10⁻³ mbar.

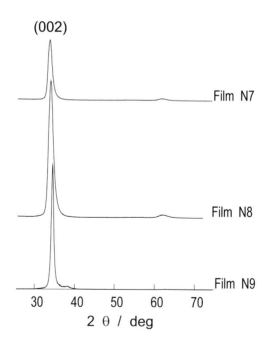

Fig. 5   XRD patterns of ZnO films prepared at a high sputtering pressure of 7.4 x 10⁻³ mbar.

spectra of the Films N7, N8 and N9, shown in Fig. 5, that under these conditions the film structure improved with increasing RF power. With a decrease of RF power, for Films N9 to N7, the (002) peak became wider and its position shifted to smaller angles.

It is evident from the results presented in Figs 2 to 5 that the higher sputtering pressure enhanced the *c*-axis orientation of the films. For a relatively low pressure of 1.8 x 10^{-3} mbar, it was found that the substrate temperature should be reduced, while the oxygen partial pressure and RF power should be altered in order to achieve the optimal growth condition. For a high sputtering pressure of 9.3 x 10^{-3} mbar, the *c*-axis orientation was improved by increasing the oxygen partial pressure (compare XRD spectra for films N1, N4 and N5 from Figs 2 and 4). In contrast, for a low sputtering pressure, this procedure led to a dramatic degradation of the XRD spectrum (Film N17, Fig. 6, oxygen partial pressure of 7.4 x 10^{-4} mbar). In this case, the effect of increasing the oxygen partial pressure was to change the orientation of the crystallites so that the *c*-axis moved from being in the main perpendicular to the substrate to being parallel to the surface, since the intensity of the reflection from the (002) planes was greatly reduced in comparison that shown in Fig. 2, and the intensity of the (110) peak significantly increased. In addition, the intensity of the reflection from the (100) planes increased. Suzuki *et al.*[12] reported analagous changes in the orientations of ZnO thin films deposited by ion-beam sputtering with increasing oxygen pressure i.e the intensity of the (002) peak decreased, while that of the (100) peak increased.

The XRD spectrum showed some improvement on decreasing the oxygen partial pressure to 3.6 x 10^{-4} mbar for the same RF power of 90 W (Film N16, Fig. 6). In this case, the

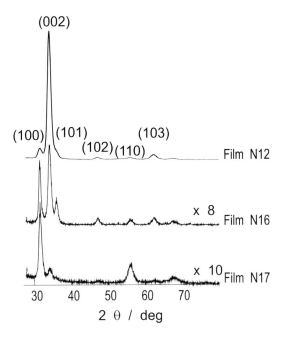

Fig. 6 XRD patterns of ZnO films prepared at a low sputtering pressure of 1.8 x 10^{-3} mbar.

(002) peak was found to dominate the spectrum. Decreasing the RF power for this oxygen partial pressure to 50 W (Film N12, Fig. 6) gave a significant improvement in the *c*-axis orientation, although (100) and several small peaks reflected from the other planes were also present in the spectrum.

The same character of the XRD spectrum as for Film N12 was achieved for Film N10, shown in Fig. 7, grown at a higher sputtering pressure of 2.4 x 10^{-3} mbar, with RF power and oxygen partial pressure equal to 50 W and 3.6 x 10^{-4} mbar respectively. Increasing the oxygen partial pressure to 4.2 x 10^{-4} mbar (Film N11, Fig. 7), did not improve the *c*-axis orientation.

The final processing parameters were repeated for sputtering ZnO films onto different substrates, including Si/SO$_2$, Si/SiO$_2$/Al and stainless steel, but the reproducibility of the XRD spectra were poor. The best oriented films deposited under these conditions of low sputtering pressure were films N10, N11 and N12. However, the (002) peaks in their XRD spectra were shifted to smaller angles and were wider than those found for the best ZnO films (Films N4–N6 and N9, Figs 4 and 5), produced at the higher sputtering pressure.

3.2 SCANNING ELECTRON MICROSCOPY

The scanning electron micrographs of ZnO films prepared under a variety of processing parameters are shown in Figs 8–11. It is evident that Film N9, fabricated at high sputtering pressures of 7.4 x 10^{-3} and with the optimum oxygen partial pressure and RF power,

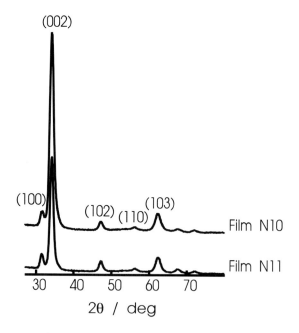

Fig. 7   XRD patterns of ZnO films prepared at a low sputtering pressure of 2.4 x 10^{-3} mbar.

Fig. 8   Surface microstructure of film N9.

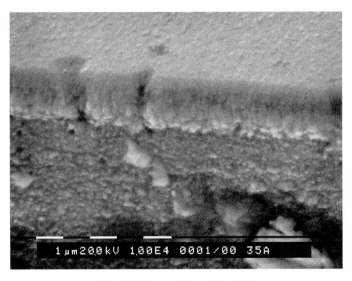

Fig. 9   Edge region of film N9.

had a regular fine-grained surface structure with a grain diameter of approximately 0.1 μm. The cross-sectional microstructure at the edge of Film N9 showed a dense columnar structure with regular grain size. At high sputtering pressure, but with high residual pressure (Film N3), the film was more irregular in structure, with an average grain diameter of approximately 0.5 mm. The microstrucrure of the Film N14, grown at the low sputtering pressure of 1.8 x 10^{-3} mbar looked regular as for Film N9 but the density was very poor.

Fig. 10   Surface microstructure of film N3.

Fig. 11   Surface microstructure of film N14.

4. DISCUSSION

The film quality depends on all of the processing parameters, including substrate temperature, sputtering pressure, argon to oxygen flow rate ratio, and RF power. The operating pressure, together with the target-substrate distance, dictate the number of collision events that the sputtered atom will have on its way to the substrate surface. The temperature of the substrate and sputtering pressure control the surface adatom mobility and the structural quality of the film.

According to the Thorton structure zone model,[13] an increase of the sputtering pressure should change the film structure from being highly dense, columnar (Zone T) to a more porous structure with a rough surface (Zone 1). At low pressures, a decrease of collision events and enhanced bombardment of the growing surface by irradiated species is expected to enhance the adatom mobility, and so nucleation can take place at a lower substrate temperature.

From a number of papers,[1-5,14] it is evident that for conventional sputtering systems, where a high-pressure atmosphere (10^{-3} mbar or higher) is required in order to generate a stable plasma, the substrate temperature must be rather high, in the region of 300–450°C. For systems incorporating an additional source or sources of activated ions, such as ion gun or microwave source, the sputtering pressure can usually be reduced to below 10^{-3} mbar, since the effect of bombardment of the substrate is more pronounced and the nucleation process can occur at lower substrate temperatures, from ambient to 200–250°C. Adatoms on the substrate surface move randomly on the substrate surface like a two-dimensional gas, so that crystal growth can occur from particular low-energy crystal planes, giving rise to preferential film orientation. Therefore, the sputtered particles must have the appropriate kinetic energy when they reach the substrate surface in order to achieve the desired film orientation.

It was shown previously[15] that for a base pressure of 1.3 x 10^{-7} mbar, an oxygen reactive gas pressure of 2.6 x 10^{-4} mbar, an operating inert gas pressure of 1.3 x 10^{-3} mbar and a target-substrate distance equal to 5 cm, the percentage of sputter ejected target atoms that arrive at the substrate surface without undergoing a collision will be approximately 32%. Increasing the operating pressure to 10^{-2} mbar reduces this quantity dramatically to about 0.0011% and in this case the sputtered atoms will have reached thermal equilibrium with the process gases and their transport to the film surface would be due to diffusion. When a sputtered atom undergoes only 2 or 3 collisions, its kinetic energy is typically more then 100 times higher than that of a thermally evaporated atom.

It is quite obvious from this discussion that in general the best quality films should be deposited at a low sputtering pressure. However some of the authors investigating the quality of magnetron sputtered films reached the conclusion that highly oriented, dense, stoichiometric films could be grown at high pressure, for example more that 6 x 10^{-3} mbar,[14] which contradicts Thorton's model. Analagous results were found from our investigation, since better control over the film structure was obtained at relatively high sputtering pressures of 7.4 x 10^{-3} and 9.3 x 10^{-3} mbar. Well oriented, dense films with good reproducibility as a function of sputtering time and for different substrates were grown at such high sputtering pressures.

One of the main reasons for this non-correspondence of the experimental results to the theory may be the rather high pressure of residual gases during the sputtering of these films, for example around 5 x 10^{-5} mbar, as reported by Meng.[14] For magnetron sputtering systems, the species of residual gas can include the following:

(i)   residual vacuum background gas (oxygen, nitrogen, water vapour, atomic and molecular radicals);
(ii) impurities in the target material (plus possibly simple poisoned oxide molecules);
(iii) impurities in the processing gases;
(iv) impurities from the target clamping arrangement and substrate holder framework.

It is clear that all of the species which can strike the film growth surface, which includes all of the plasma processing species (argon sputtering gas atoms/ions, oxygen reactive gas atoms/ions, zinc target atoms) and all species of residual gas, can influence the overall film quality and performance.

When the sputtering pressure in our experiments was relatively high, equal to 7.4 x 10^{-3} or 9.4 x 10^{-3} mbar, all species in the chamber lose their energy due to collisions and the ratio of residual gas species to the plasma processing species is smaller. In this case, the species of the residual gas do not influence the quality of the films as strongly as in the case when the sputtering pressure is low, around 1.8 x 10^{-3} or 2.4 x 10^{-3} mbar. The number of collisions in the latter case is smaller, and more species with higher energy strike the growing surface. Therefore, it is apparent from our results that at low pressures the temperature required for the nucleation process was smaller than that of films prepared at the higher sputtering pressure (250 °C compared with 300 °C). The ratio of the residual gas species to the processing gas species and target atoms is higher for a low sputtering pressure, and all of the residual species disturb the growing process in an uncontrolled way. Furthermore, some of these species could be incorporated in the growing film. The direct consequence of this effect is poor reproducibility of the film structural quality. The films obtained at low pressures were porous, in compressive stress, with mixed crystallite orientation and with a tendency for the *c*-axis to be parallel to the substrate when the oxygen partial pressure was increased.

It was reported previously[16] that films grown at higher sputtering pressures (around 10^{-2} mbar) also exhibit lower residual stress. Because the value of the SAW coupling coefficient was only slightly reduced at this pressure, it is possible that films produced under the same processing parameters as those for films N4, N5, N6 or N9 can be successfully used for SAW applications.

The deposition rate in our experiments increased rapidly as the RF power increased or the oxygen partial pressure decreased. A more gradual increase in deposition rate was observed as the sputtering pressure was decreased. Typical deposition rates were in the range 1.5–3.2 µm h^{-1}. The partial pressure of the reactive oxygen together with RF power directly affected the chemical stoichiometry and homogeneity of the film. The influence of the RF power level on the film properties is presumably caused by a change in the generation rate of active species. It is evident from our experiments that the stress in compound films such as ZnO depends not only on the sputtering pressure[16] or the sputtering pressure together with substrate-target distance (*Pd* product), as was reported for metallic coatings.[17] The shift in the (002) peak due to strain was found to depend on the ratio of oxygen partial pressure to power supply as well, and in fact on the rate of the deposition. The use of higher oxygen partial pressures and higher operating pressures led to the growth of highly oriented, less stressed films. However, the same oxygen level at low pressure caused a dramatic worsening of the film structure to occur.

Preliminary investigations of the film resistivity gave values of around $10^4\,\Omega$m for the films N11 and N12, which were grown at low sputtering pressure. The resistivities of the films produced at high sputtering pressures was greater than $10^5\,\Omega$m, which indicates that their stoichiometry is better, because a low resistivity is usually associated with a considerable oxygen deficiency and/or the concentration of interstitial zinc ions.[11]

5. CONCLUSIONS

ZnO films were deposited by RF reactive magnetron sputtering in order to identify processing parameters suitable for producing films for SAW resonator applications. In the cases where the films were found to exhibit a high degree of orientation, the deposition process was repeated under nominally identical conditions onto Si/SiO$_2$, Si/SiO$_2$/Al and stainless steel substrates. For high sputtering pressures of 7.4 x 10^{-3} and 9.3 x 10^{-3} mbar, the substrate temperature required to achieve high quality films was relatively high, around 300 °C. At the optimum ratio of oxygen partial pressure to RF power, the ZnO films exhibited a dense, smooth surface, strain free, fine grained structure of highly oriented crystallites with the *c*-axis perpendicular to the substrate. The films were deposited with good reproducibility of the structural quality onto the various substrates under investigation. At low sputtering pressures of 1.8 x 10^{-3} and 2.4 x 10^{-3} mbar, the substrate temperature suitable for the optimum nucleation process was found to be lower and equal to 250 °C. Films deposited under these conditions were porous, in compressive stress, with mixed crystallite orientation and a tendency to have the *c*-axis parallel to substrate when the oxygen partial pressure was increased. This non-correspondence to Thorton's zone model can explained by a rather high level of the residual pressure, which disturbs the growing process at low sputtering pressures.

ACKNOWLEDGMENTS

The work reported here was carried out as part of the EPSRC research project 'Piezoelectric Thin Films for SAW Strain Sensors', grant no. GR/J37401. The authors would like to thank EPSRC for providing financial support.

REFERENCES

1. B.T. Khuri-Yakub and J.G. Smitt: 'Reactive magnetron sputtering of ZnO', *IEEE Ultrason., Symp. Proc.*, 1980, 801–804.
2. T. Yamamoto, T. Shiosaki and A. Kawabata: 'Characterization of ZnO piezoelectric films prepared by RF planar-magnetron sputtering', *J. Appl. Phys.*, June 1980, **51**(6), 3113–3120.

3. G. Perluzzo, C.K. Jen and E.L. Adler: 'Characteristics of reactive magnetron sputtered ZnO films', *IEEE Ultrason. Symp. Proc.*, 1989, 373–376.
4. S. Tacite: 'Comparison between optical property and crystallinity of ZnO thin films prepared by RF magnetron sputtering', *J. Appl. Phys.,* 15 May 1993, **73**(10), 4739–4742.
5. M. Kadota, T. Kashimani and M. Manakata: 'Deposition and piezoelectric characteristics of ZnO films by using an ECR sputtering system', *IEEE Transaction on Ultrasonics, Ferroelectrics and Frequency Control,* 15 May 1993, **41**(4), 479–483.
6. A.L. Nalamwar and M. Epstein: 'Effects of applied strain in ZnO thin-film SAW devices', *IEEE Transactions on Sonics and Ultrasonics,* May 1976, **SU-23**(3), 144–147.
7. J.H. Visser, M.J. Vellekoop, A. Venema *et al*: 'Surface acoustic wave filters in ZnO-SiO$_2$-Si layered structure', *IEEE Ultrason. Symp. Proc.*, 1989, 195–200.
8. S.J. Martin, R.L. Gunshor and R.F. Pierretp: 'High Q, temperature stable ZnO-on-silicon SAW resonator', *IEEE Ultrason. Symp. Proc.*, 1980, 113–117.
9. S. Ono, K. Wasa and S. Hayakawa: 'Surface-acoustic-wave properties in ZnO-SiO$_2$-Si layered structure', *Wave Electronics*, 1977, **3**, 35–39.
10. R.L. Gunshor: 'The interaction between semiconductors and acoustic surface wave – a review', *Solid State Electronics*, 1975, **18**, 1089.
11. W. Hirschwald, P. Bonasewicz and *et al*: 'Zinc oxide: Properties and behaviour of the bulk, the solid (vacuum and solid) gas interface', *Current Topics in Materials Science,* North-Holland, Amsterdam, 1981, **7**, 143–482.
12. Y. Suzuki, T. Yotsuya, K. Takiguchi, M. Yoshitakey and S. Ogawa: 'The effect of charged particles when preparing ZnO thin films by ion beam sputtering deposition', *Appl. Surf. Sci.*, 1988, **33/34**, 1114.
13. J.A. Thornton: 'Influence of apparatus geometry and deposition conditions on the structure and topography of thick sputtered coatings', *J. Vac. Sci. Technol,* 1974, **11**, 666–670.
14. L.-J. Meng and M.P. dos Santos: 'Direct current reactive magnetron sputtered zinc oxide thin films – the effect of the sputtering pressure', *Thin Solid Films*, 1994, **250**, 26–32.
15. V.P. Kutepova, D.A. Hall, and S.P. Speakman: 'Piezoelectric thin films for SAW devices', *Ceramic Films and Coating,* British Ceramics Proceedings 54, The Institute of Materials, 1995, 189–205.
16. J.C. Zesch, B. Hadimioglu, B.T. Khuri-Yakub and *et al*: 'Deposition of highly oriented low-stress ZnO films', *IEEE Ultrason. Symp. Proc.*, 1991, 445–448.
17. C. Hudson, R.E. Somekhs: 'Stress in UHV planar magnetron sputtered films', *Mat. Res. Soc. Symp. Proc.*, Material Research Society, 1992, **239**, 144–150.

DEFECT DETECTION IN RODS USING PZT SENSORS

K.T. FEROZ AND S.O. OYADIJI

Dynamics and Control Research Group, Division of Mechanical Engineering, Manchester School of Engineering, University of Manchester, Manchester M13 9PL, U.K.

ABSTRACT

PZT sensors are used to detects defect in cylindrical rods using wave propagation techniques. Stress pulses are measured and predicted using the finite element method in rods free of defects and with a defect. The defect is introduced in the rod in the form of a small slot. By analysing the stress wave data for the defect free rod and for the rod with a defect, it is possible to pinpoint the location of the defect. PZT tiles of dimensions 5 x 3 mm, which are cut from standard PZT patches of dimensions 30 x 30 mm, are bonded to the surface of the cylindrical rods and are used for monitoring the propagation of stress pulses induced in the rods by the collinear impact of spherical balls on one of the plane ends of the rods. A finite element analysis is performed to predict the stress pulses in the rods. The results show that the defect can be located using this technique. It is shown that a high degree of correlation is obtained between measured and predicted characteristics.

1. INTRODUCTION

The field of non-destructive testing makes wide use of ultrasonic and wave propagation techniques to detect defects in structures and machinery. This is a very important area of research since the early detection and location of defects is paramount to the prevention of catastrophic failures. Some of the defects, such as cracks and voids, are often difficult to detect visually because they are either too small to be seen by the naked eye or they occur in inaccessible parts of a structure or structural component. In classical, ultrasonic techniques, pairs of ultrasonic transducers, which comprise an emitter and a receiver, and are made of quartz crystals are used.

Several investigations have been reported in the literature on wave propagation studies[1,2] and on defect detection using vibration and wave propagation techniques.[3-6] In the work reported here, PZT sensors are used in conjunction with longitudinal wave propagation phenomena in a solid rod for detecting a defect in the rod. The defect is introduced in the rod in the form of a slot while a longitudinal wave is induced in the rod via the collinear impact of a spherical ball on a plane end of the rod.

Traditionally, stress wave propagation in solids is monitored by means of stainless steel gauges. However, the output signals of the gauges are very small and they therefore require amplification. The general approach for providing amplification is the use of a Wheatstone bridge circuit and an additional amplifier system. But because the frequency content of the stress pulses sometimes extends into the ultrasonic range, it is important that the amplifiers used have a very wide frequency bandwidth. Thus, for experiments in which a large number of strain sensors are required for detecting high frequency stress waves, the use of stainless steel gauges requires a number of expensive amplification systems.

Conversely, PZT tiles of the same physical dimensions as the conventional stainless steel gauges provide voltage outputs, when deformed or stressed by a propagating wave, that are greater than the outputs of the conventional gauges by factors up to 10 000 or more. As a result of this, PZT tiles produce signals that are readily measurable without the need for amplification. This makes the use of PZT tiles for stress wave propagation measurements very attractive. Thus, stress wave propagation experiments which require a large number of stress/strain sensors can be easily and cheaply implemented using PZT tiles. However, there is the need to determine the gauge factors of the PZT tiles when used for wave propagation measurements. This calibration process is carried out in the present work by means of the finite element method (FEM).

In this paper, the propagation of stress pulses in rods with and without a defect is investigated experimentally and numerically. PZT sensors were cemented to the surface of the rods at different axial locations. A time domain FE simulation of wave propagation in rods was performed using the commercial FE code ABAQUS.[7] The results show that a defect can be located using this technique. It is shown that a high degree of correlation is obtained between measured and predicted characteristics.

2. CALIBRATION OF PZT SENSORS USING FEA DATA

The PZT sensors used in the present work were of dimensions 5 x 3 mm and were cut from standard PZT patches of dimensions 30 x 30 mm using a diamond disc cutter. In order to calibrate these sensors, a random selection of pairs of these sensors were bonded at 5 locations along the length of a 10 mm diameter by 2.44 m long steel rod. At each location, 2 PZT sensors were bonded at diametrically opposite positions as shown in Figs 1 and 2. They were so arranged in order to minimise the effects of bending waves produced by eccentric impact. The rod was subjected to transient deformation via the collinear impact of spherical balls of diameters of 15 mm and 25 mm. The stress pulses monitored by the PZT sensors were recorded using a 4 channel 50 MHz transient recorder. A finite element analysis (FEA) of the propagation of longitudinal stress waves in the rod was then performed using the ABAQUS FE package as described in previous work.[8]

Figure 3a shows the predicted strain pulses in the rod at element position 3. The corresponding measured strain pulses at gauge position 3 are shown in Fig. 3b with the strain values being expressed in volts. By correlating the measured and predicted characteristics, the conversion factors for the PZT sensors were found to be 13.89 mstrain/Volt and 14.10 mstrain/Volt when the diameter of the impacting ball was 15 and 25 mm respec-

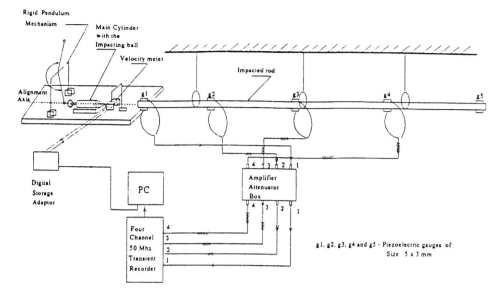

Fig. 1 Experimental set up for propagation of pulses in rods

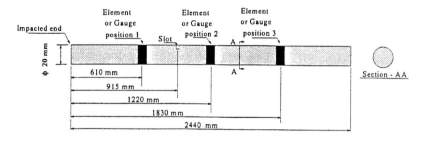

Fig. 2 Slot location on the rod and the locations of element or gauge positions for which the results are plotted.

tively. Thus, the average conversion factor of the PZT sensors is 14.0 mstrain/Volt. Using this conversion factor in conjunction with the measured strain pulses enables their direct comparison with the predicted characteristics of the rod. Figure 3c shows that the predicted characteristics correlate well with the measured characteristics.

3. FINITE ELEMENT SIMULATION OF SLOTTED AND UNSLOTTED ROD

The finite element tools ABAQUS and PATRAN were used to simulate the slot in a 20 mm diameter rod. First the analysis was performed on a mild steel, defect free rod of 2.44 m long. The rod was discretised using a combination of 15–noded and 20–noded solid

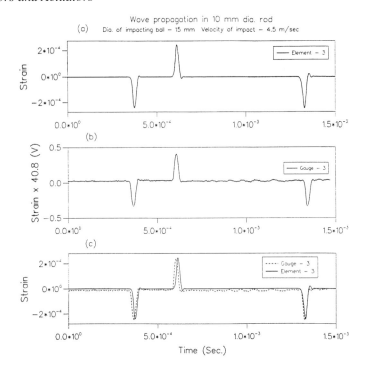

Fig. 3   Predicted and measured wave propagation in 10mm diameter defect free rod.

elements of type C3D15 and C3D20 respectively. The model had a free boundary condition. Strain pulses were induced in the bar by the collinear impact of a spherical ball.
The force-time relationship was calculated using Hertzian theory of contact. The force
spectrum evaluated from Hertzian theory of contact was then used in the FE code
ABAQUS as an input force of impact using the *AMPLITUDE option. This force was
applied at the centre of the input end of the rod. Figure 2 shows the element and sensor
positions of interest.

The slot was introduced at the location of interest as shown in Fig. 2 by providing a
discontinuity or gap between the elements. The dimensions of the gap represented the
size of the slot. A combination of 2D and 3D elements were used for the discretisation.
The region of the rod encompassing the slot was modelled with 3D solid elements C3D20
and C3D15 while all the other regions were modelled with 2D axisymmetric CAX8 solid
elements. Figure 4(a) shows the predicted pulses in a 20 mm diameter defect free rod at
chosen element positions. Strain pulses for the rod with a slot are shown in Fig. 4(b). The
smaller pulses between the bigger pulses are due to the slot.

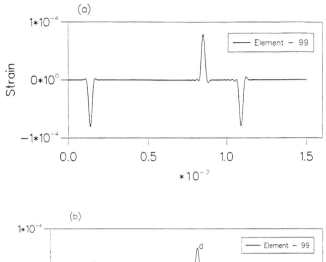

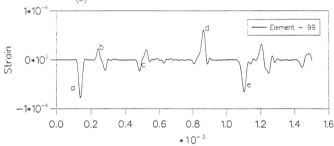

Fig. 4   Predicted wave propagation in 20 mmdiameter rod:
(a) defect free rod, (b) rod with a 2 x 6 mm slot

4. EXPERIMENTAL MEASUREMENT OF STRESS WAVES IN THE SLOTTED AND UNSLOTTED ROD

Experiments were performed on the 20 mm diameter rod with and without a slot. A slot of dimensions 2 mm wide by 6 mm deep was cut in the rod at a location of 915 mm from the impacted end using a milling machine. The piezoceramic PZT sensors were cut out of 30 mm x 30 mm square pieces to sizes of 5 mm x 3 mm and were cemented to the surface of the rod. They were located at five positions along the length of the rod as shown in Fig. 1 which also show the overall experimental set-up. The signals monitored by these gauges were passed through a four channel attenuator, and were sampled and stored in the four channel 50 MHz transient recorder card which was installed inside a computer. A rigid pendulum mechanism was used to impact the spherical balls, which in turn, impacted a plane end of the rods in the longitudinal direction.

The measured longitudinal strain pulses for the rod free of defect and with a defect are shown in Fig. 5. The gauge positions 1, 2 and 3 correspond to element numbers 99 , 61 and 31 respectively. The Y-axis of these figures represent the amplitude of the pulses in volts multiplied by the attenuation factor. The presence of the slot is indicated by the pulses of smaller peak amplitudes (pulses b, c, etc.), in between the bigger pulses (pulses

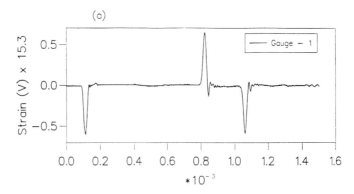

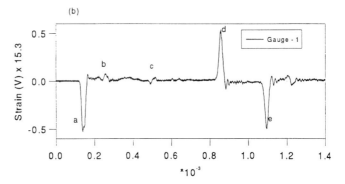

Fig. 5 Measured wave propagation in 20 mm diameter rod:
(a) defect free, (b) rod with a 2 x 6 mm slot located at x = 915 mm.

a, d, etc.). The procedure for estimating the location of the slot from this data is illustrated in section 5.

5. SUMMARY OF THE RESULTS

A comparison of predicted and measured pulses obtained after multiplying the measured pulses in volts by the conversion factor is shown in Fig. 6. It is obvious that there is reasonable agreement between the measured and predicted characteristics. The results demonstrates the effectiveness of PZT sensors for monitoring stress pulses in structures subjected to impact loading.

Figure 4b shows the propagation of pulses in the 20 mm diameter rod with a slot at a distance 915 mm from the impacted end. Three element positions on the rod were chosen. These elements were spaced equally, at a distance of 610 mm. The spacing of the elements was related to the gauge positions on the rods for the experimental study. Waves propagated along the rods from the impacted end to the other end. Longitudinal strain pulses were plotted against the time for a period of few microseconds.

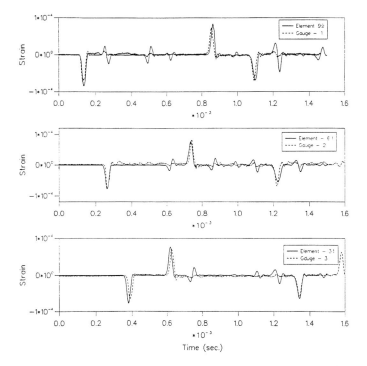

Fig. 6 Comparison of FE data with the experimental data for the 20 mm diameter rod with a 2 x 6 mm slot.

The first trough 'a' in Fig. 4b represents the primary compressive wave. This is the point at which the primary compressive pulse reaches the particular element position (in this case element 99, which is at a distance of 610 mm from the impacted end). The pulse then travels to the slot where it is reflected and returned as a secondary tensile pulse, which is small in amplitude and represented by the first peak 'b'. This secondary pulse travels to the impacted end, gets reflected as a secondary compression wave which is represented by the trough 'c'.

Meanwhile, the primary compressive pulse also travels to the other end of the rod where it is returned and reflected as a primary tensile pulse represented by peak 'd'. This primary tensile pulse then travels to the impacted end, gets reflected and comes back as a primary compressive pulse (trough 'e'). The pulses 'b' and 'c' are due to the slot. By studying the propagated pulses at different element positions, the location of the slot can be predicted.

It can be shown that the distance D_c from a reference element (or gauge) position to the slot is given by

$$D_c = \frac{t_2 - t_1}{2} C_0 \qquad (1)$$

where t_1 is the time taken by the pulse to travel from the impacted end to the reference element (gauge) position, t_2 is the time taken by the pulse to travel from the reference element to the slot and back to the element, and C_o is the wave velocity in the rod.

From Fig. 4b for element position 1, which corresponds to finite element number 99, t_1 = 125 ms and t_2 = 250 ms. For mild steel, $C_o \approx 5000$ ms^{-1}. Therefore $D_c = 0.3125$ m. Thus, the calculated distance of the slot from the impact end = $0.610 + D_c = 0.9125$ m. The actual distance is 0.915 m. Hence, the estimate of the slot location is in error by 0.8%.

6. CONCLUSIONS

A procedure has been developed to detect defects in a rod using experimental and finite element techniques. The experimental results show that PZT sensors can be used successfully for detecting defects in a rod. The average conversion factor of the PZT was found to be 14.0 mstrain/Volt. A high degree of correlation is obtained between the measured and predicted pulses. FE analysis, using the combination of two and three dimensional elements, is found to be a very effective and faster method for problems like slot detection in large models. It has been demonstrated that the FE simulation can play an important role in predicting the wave phenomena and help in the planning of non-destructive testing (NDT) experiments.

ACKNOWLEDGEMENTS

The authors wish to thank Mr Tony Taylor, Senior Experimental Officer, and Mr Steve Longden, Experimental Officer, for the design and implementation of the four channel transient recorder card.

REFERENCES

1. M.M. A-Mousawi: 'On Experimental Studies of Longitudinal and Flexural Wave Propagation: An Annotated Bibliography', *Applied Mechanics Review*, 1986, **39**(6), 853–863.
2. C.S. Barton, E.G. Volterra and S.J. Citron: 'On Elastic Impacts of Spheres on Long Rods', *Proc 3rd US Natl Cong Appl Mech*, 1958, 89–94.
3. G. Hearn and R.B. Testa: 'Modal Analysis For Damage Detection in Structures', *Journal of Structural Engineering*, 1991, **117**(10).
4. H. Baruh and S. Ratan: 'Damage Detection in Flexible Structures', *Journal of Sound and Vibration*, 1993, **166**(1), 21–30.
5. P. Cawley and R. Ray: 'A Comparison of the Natural Frequency Changes Produced by Cracks and Slots', *Transactions of the ASME*, 1988, **110**, 366–370.
6. G.L. Qian, S. N. Gu and J. S. Jiang: 'The Dynamic Behaviour and Crack Detection of a Beam with a Crack', *Journal of Sound and Vibration*, 1990, **138**(2), 233–243.

7. ABAQUS and PATRAN Manual.
8. K. T. Feroz and S. O. Oyadiji: 'Use of PZT Sensors for Wave Propagation Studies', *Smart Structures and Materials 1996: Smart Sensing, Processing, and Instrumentation*, K. A. Murphy and D. R. Huston eds, 1996, Proc SPIE 2718, 36–46.

HOW TO CHOOSE AN ACTUATION MECHANISM IN A MICROENGINEERED DEVICE

D. WOOD*, J.S. BURDE[†] AND A.J. HARRIS

*School of Engineering, University of Durham, South Road, Durham, DH1 3LE
[†]Dept. of Mechanical Engineering, University of Newcastle-upon-Tyne, Stephenson Building, Newcastle-upon-Tyne, NE1 8ST
Dept. of Electrical and Electronic Engineering, University of Newcastle-upon-Tyne, Merz Court, Newcastle-upon-Tyne, NE1 7RU

ABSTRACT

The subject of mechanical actuation in microengineered structures is an expanding area of study. Amongst the easiest actuation mechanisms to understand are capacitive, electromagnetic and thermal. Capacitive techniques are material independent, but require very small separations to be effective. Electromagnetic actuation is similarly material independent, but does need an external magnetic field. Typical thermal actuators are either bimetallic in form, or incorporate diffused resistors to dissipate the heat.

A lot of recent research has been generated in using material dependent properties such as piezoelectricity, the shape memory effect, electrostriction and magnetostriction. These are more developed in large structures, but on the micro scale the reproduction of material dependent properties has been a major problem limiting the application of techniques that promise a large actuation force and simple fabrication procedures.

INTRODUCTION

Many factors need to be considered in microengineered actuators. These include:

- ease of understanding of the mechanism,
- predictability and modelling,
- a material effect, like piezoelectricity, vs material independent, like electrostatic,
- a consideration of new materials – characterisation and new processes, reproducibility, compatibility with the existing fabrication line, customer acceptability, manufacturing yield and device reliability then become important,
- the size of the actuation force, and how much actuation benefit is produced when the forces are scaled to the size of your device,
- the fractional stroke (displacement per unit length) of the actuator,
- any temperature dependence,
- the frequency response.

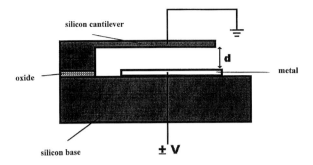

Fig. 1 A capacitively actuated cantilever.

This paper will introduce all the main mechanisms currently used in microengineering, and will try and address the above issues in relation to each technique.

CAPACITIVE

Electrostatic drive systems have a strong position in silicon microengineering. The arguments put forward for capacitive actuation are familiar: easy to understand, readily calculable, compatible with CMOS processes and materials, little temperature dependence and a force that scales well to the micro dimension.

In every application two electrodes are separated by an air gap. One electrode is usually fixed, and the other is free to move. An example of a cantilever is shown in Fig. 1. Because of the square dependence of the electrostatic force, making the air gap narrow is very advantageous: typical values are in the 2–5 µm region. This type of structure, and the dimensions involved, are standard to microengineering technology; however, careful techniques have to be applied here to prevent the problem of 'stiction' of the electrodes. The electrode gap should always be in a vacuum (unless the actuator is a pump) because energy will be dissipated in moving the air in and out of the electrode spacing.

Capacitive actuation is material independent, and purely a dimension and voltage effect. In addition, as an electrode deflects, the separation decreases and the force increases. The electrode separation will tend to become spatially non-uniform, as shown in Fig. 2. Hence calculations of the forces involved will need time and distance–dependent integration. Also, an electrostatic force will always attract, and so cannot be used, for instance, to increase the electrode spacing.

A further advantage to the exploitation of electrostatic actuation can be obtained via the Paschen effect. The dielectric breakdown field strength in air is around 3×10^6 Vm^{-1}, but when the gap between two electrodes falls below about 3 µm the breakdown field strength increases dramatically, as can be seen from Fig. 3. Thus very large driving voltages can be used to increase the actuation force.

Because of the attractive nature of the force a free electrode will keep bending until it

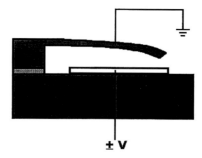

Fig. 2   Bending of the cantilever under an attractive force.

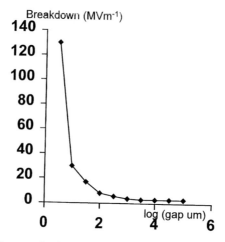

Fig. 3   Increase in electrostatic breakdown field with decreasing
electrode gap (the Paschen effect).

is held in balance by an axial tensile load. However, if the drive voltage exceeds a certain value called the pull-in voltage, then there is a rapid build up of the electrostatic forces, and the system becomes unstable, as in Fig. 4. Care should be taken to avoid this.

Another point to note regarding this type of actuation is that electric fields interact with most materials, meaning that electrostatic actuators may need greater environmental isolation than other techniques. Also, the unwanted ability of electrostatic fields to attract dust, and their adverse effect on CMOS circuitry, are well known.

ELECTROMAGNETIC

A common form of magnetic actuation is based on the Lorentz effect. This method is ideally suited for large electromagnetic applications, but scales poorly to microengineering dimensions. Magnetic actuation becomes quite weak on a microscale, unlike capacitive

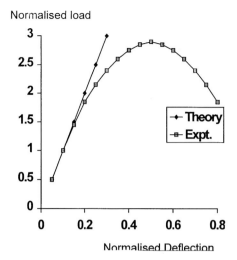

Normalised load

Fig. 4 Instability of capacitive actuation beyond the pull-in voltage.

actuation. However, it has been used to resonate one of our gyroscope structures, as in Fig. 5(a). There is a conductor in the 'top right' that passes current over a long length of silicon. When a magnetic field is applied perpendicular to the plane of the device, the silicon length moves at right angles to field and current according to Fleming's left hand rule. This movement acts on the connecting arm, which has its other end located off centre of the gyro. Now, there is a similar structure in the 'bottom left' of the gyro. If an ac current is applied at the device's resonant frequency, and the current is arranged to move the connecting arms in anti-parallel directions, the bow-tie will resonate in the plane of the device – as in the finite element plot of Fig. 5(b).

Many devices incorporate a solenoid to increase the level of actuation. Although such structures are straightforward to make on a macroscale, in microengineering they involve a lot of processing steps to enable the conductor and magnet to be wrapped around each other. Several laboratories have reported solenoid actuated structures,[1] including micromotors.

Material issues play a part: the flux density of the permanent magnet and the current carrying capability of the conductors need to be maximised. For the latter parameter, the current densities can be much higher than for large electromagnetic structures, because the increased surface area/volume ratio means that small conductors can more readily dissipate heat. However, the passing of a current through a conductor will cause resistive heating – this is the basis of thermal actuation (see below). If the current density is high, the thermal term may compete with the electromagnetic and cause unwanted actuation.

Actuation control is via current rather than voltage: this means that electromagnetic devices may be more suitable for bipolar rather than CMOS integration. A large force means a high current, leading to considerations of power consumption and dissipation. The need for external magnets is a cost disadvantage, and the additional processing

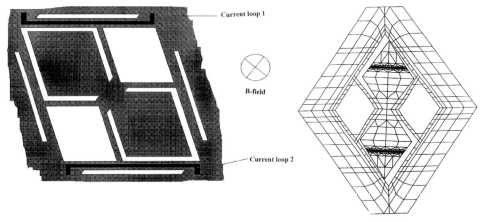

Fig. 5   (a) A Lorentz force actuated gyroscope. (b) In-plane resonant motion of the device.

required for the production of a solenoid is in contrast to the simplicity of electrostatic actuation. But only low voltages are needed, which is an advantage when it comes to battery operation (however, the power consumption is a disadvantage in this respect). In addition, magnetic actuation does not suffer catastrophic field breakdown and is tolerant to dust, humidity and surface topography. It has one other distinguishing feature over the electrostatic force: by reversing the current direction, movement can be made to occur in two opposite directions, as used in the gyroscope of Fig. 5.

PIEZOELECTRIC

Piezoelectricity is a property of insulating materials where an applied voltage generates an external stress: the mechanism is reversible. The force produced is quite large, the response times short, and efficiencies are high if the leakage currents and hysteresis effects are kept under control. However, the fractional stroke (displacement per unit length) is quite small, typically being no more than 0.2%, and is direction dependent.

Increasing use is being made of titanate materials, in particular PZT (lead zirconate titanate) because of the high value of the d_{31} piezoelectric coefficient. This material is very popular because of its ability to be deposited from a solution (the sol-gel process). The great interest in sol-gel has come from the fact that a planar structure can be achieved without the need for air gaps between electrodes. The sol-gel process typically starts with the synthesis of a precursor solution which contain the active materials; the precursor liquids can then be deposited on a substrate by spin coating. The material on the substrate is then heated and poled before being allowed to cool back to room temperature.

Piezoelectricity is a phenomenon that is highly material dependent. Problems remain in turning theoretical predictions into reliable applications, most notably in the uniform-

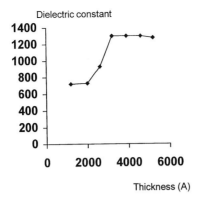

Fig. 6 Typical change in the dielectric constant for a thin film piezoelectric material.

ity of material deposited via the sol-gel process. It has proved difficult to transfer bulk material properties to thin film form: problems occur with grain size, in-built stress and crystal orientation. Typical results of dielectric constant vs film thickness are shown in Fig. 6. An additional issue is that the piezoelectric coefficients are temperature as well as direction dependent, and are often dependent on the operating frequency.

THERMAL

Heat applied to the top of a bimorph beam will diffuse down through the structure to create a temperature gradient. Differential thermal expansion of the material will then introduce a mechanical moment to bend the structure. Because of the high value of the expansion coefficient of aluminium, this metal has been a popular choice for metal-silicon bimorph structures: an example of such a device in shown in Fig. 7.

This is a very simple actuation mechanism, which can be incorporated quite readily into any structure. Drive voltages are low and large deflections can be produced, although by their very nature thermal actuators consume a lot of power. For many structures such as a cantilever thermal actuation can be preferred over electrostatic because the movement of the structure does not subsequently affect the actuation force. It will be noted, however, that analysis is not especially straightforward: consideration of power dissipation in the materials, heat conduction between the layers, heat loss radiated from the structure, and the change in material properties as the temperature rises must all be taken into account.

For practical devices, the dynamic behaviour of thermal excitation will always be a problem, because of the need to dissipate the residual heat from the previous cycle before the next can be implemented: hence operating frequencies tend to be low compared to other mechanisms. However, this is not as bad as may be first thought, for similar reasons to the electromagnetic actuation. The surface area/volume ratio of the heating

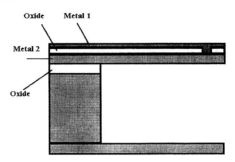

Fig. 7   A thermally actuated cantilever.

elements is favourable with small dimensions, and hence drive frequencies in the kHz region can be achieved.

Thermal actuators are limited to out of plane motion: it would be difficult to make an effective bimorph that would work in the plane of a silicon wafer. A further issue unique to thermal actuation is that a bimorph will tend to deflect under atmospheric temperature gradients. A last point is that thermal actuation force scales badly. For a given heating element, the current goes down as the square of dimension (as in the case of magnetic actuation above), whilst the resistance increases linearly: this gives a cubic reduction in heat dissipation with dimension.

SHAPE MEMORY ALLOY (SMA)

The shape memory effect occurs in certain alloy materials such as TiNi. SMAs are plastically deformable materials that, on heating to the appropriate temperature, change their crystal structure: for TiNi this is a change from the martensite to the austenite phase, which has a higher symmetry. The typical change in material properties[2] can be seen from Fig. 8.

Again this is a material property, with the associated problems of reproduction in thin film form. Thin film deposition methods can be used, but usually a high temperature (>500 °C) crystallisation schedule is needed after deposition. However, research results are beginning to show encouraging results when using shape memory effects with microengineered devices.[3]

TiNi is corrosion resistant, has good high temperature stability and an acceptable resistance. Frequency responses are slow, with heat dissipation limitations similar to those on thermal actuation. A more fundamental limit comes from the heat driven phase change nature of the shape memory effect, as such it will be limited by the efficiency associated with the Carnot cycle at 5% or less.

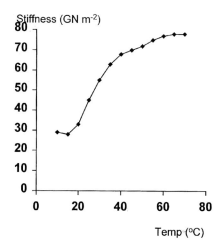

Fig. 8   The change in material stiffness as a result of the phase
transition in a shape memory alloy.

OTHER MECHANISMS

There are many other actuation mechanisms that have been used in microengineering: a
selection of these are mentioned here. They include magnetostriction,[4] electrostriction,[5]
travelling wave induced electroconvection,[6] boiling a liquid in a sealed cavity[7] and in-
voking an electrochemical reaction.[8] All are finding increased interest as an actuator
mechanism.

CONCLUSIONS

An attempt to produce an all-encompassing comparison of actuation types would be
misleading. As can be seen from the discussion in this paper, there are many different
types of actuation mechanisms to choose from and use in microengineering. None has
dominance over any other, and the choice of mechanism is a balance between many
different factors, only some of which are related to the actuation benefit. Some are better
than others in certain devices, some are mature, some promise greater benefit: the overall field is not sufficiently advanced to say whether new materials can be controlled and
characterised to realise their full potential. Difficulties in material properties, reproducibility and process compatibility are all key issues.

REFERENCES

1.   C.H. Ahn and M.G. Allen: 'A Fully Integrated Micromagnetic Actuator with a
     Multilevel Meander Magnetic Core', *IEEE Solid-State Sensor and Actuator Work-*

shop, Technical Digest, IEEE, June, 1992, **#92TH0403-6**, 14–18.
2. I.W. Hunter and S. Lafontaine: 'A Comparison of Muscle with Artificial Actuators', *IEEE Solid-State Sensor and Actuator Workshop,* Technical Digest, IEEE, June, 1992, **#92T110403-6**, 178–185.
3. R.H. Wolf and A.H. Heuer: 'TiNi (Shape Memory) Films on Silicon for MEMS Applications', *IEEE J Microelectromechanical Systems,* 1995, **4**(4), 206–212.
4. M.R.J. Gibbs: 'Piezomagnetic Control of Machines', *lEE Colloquium 'Active Drives for Microengineering Applications',* Digest 1995/**085**, 5/1–5/3.
5. R. Kornbluh, R. Pelrine and J. Joseph: 'Elastomeric Dielectric Artificial Muscle Actuators for Small Robots', *Proc. 3rd IASTED Int. Conf on Robotics and Manufacturing,* Cancun, Mexico, 14–16 June, 1995.
6. G. Fuhr, R. Hagedorn, T. Muller, W. Benecke and B. Wagner: 'Pumping of Water Solutions in Microfabricated Electrohydrodynamic Systems', *Proc. IEEE Micro Electro Mechanical Systems Workshop,* Travemunde, Germany, 4–7 February, 1992, 25–30.
7. P.L. Bergstrom, J. Ji, Y.-N. Liu, M. Kaviary and K.D. Wise: 'Thermally Driven Phase-Change Microactuation', *IEEE J. Microelectromechanical Systems,* 1995, **4**(1), 10–17.
8. C.R. Neagu, J.G.E. Gardiniers, M. Elwenspoek and J.J. Kelly: 'An Electrochemical Microactuator: Principles and First Results', *IEEE J Microelectromechanical Systems,* 1996, **5**(1), 2–9.

HIGH POWER LIGA WOBBLE MOTOR WITH INTEGRATED SYNCHRONOUS CONTROL

V.D. SAMPER*, A.J. SANGSTER, R.L. REUBEN,
K. SHEA, S.J. YANG AND U. WALLRABE†

*Department of Computing and Electrical Engineering, Heriot Watt University,
Edinburgh, EH14 4AS*

ABSTRACT

Enhanced power versions of millimetre scale electrostatic wobble motors, developing gross motive torques of approximately 1 µNm, are presently under evaluation. To improve the output power to volume ratio of motors of this type, the surface area contained within the motor envelope has been maximised. A twin stator design has been fabricated to investigate the effect of concentric surface pairs. Synchronous control of the excitation signals is necessary to improve the operational efficiency, since asynchronous actuation disregards any loading or inertial effects on the true dynamic performance. Rotor positional information, necessary for synchronous control, is provided by segmenting the bearing, where the momentary continuity between adjacent segments is detected as the rotor rolls around the perimeter of the bearing. The prototype devices are 300 microns tall, with the rotor and stator fabricated separately and assembled manually. The stator outside diameter is 2600 microns and gear ratios of 225 and 113 and minimum rotor-stator air gaps of 5 and 10 microns have been produced. A complete dynamic model of the motor, incorporating analytic and finite element field predictions, has also been developed to determine the dynamic characteristics of the wobble motor.

1. INTRODUCTION

The therapy of atherosclerosis, a type of cardiovascular disease has, in recent years, involved minimally invasive treatment of the arterial plaque found in diseased arteries. This does not require general anaesthesia or a chest incision and has the additional advantage over bypass surgery of reduced cost and decreased hospital and recuperation times.[1, 2]

To date, the most popular such alternative to bypass surgery for the treatment of atherosclerosis has been that of balloon angioplasty,[3–5] which uses a pressurised balloon to dilate the diseased section of artery and increase the through flow of blood without plaque removal. The procedure is, however, less suitable for the treatment of non-regular plaques (e.g. heavily calcified, eccentric or total occlusions) and has several known limitations, including the risk of abrupt closure, i.e. sudden artery collapse immediately following treatment, and a high rate of restenosis (regrowth of plaque tissue) during the following months.[3–5]

* now at Institute of Microelectronics, Singapore Science Park II, Singapore 117684
† now at FZK, D–76021 Karlsruhe, Germany

More recently, several devices have been introduced which are capable of removing plaque from the arterial wall. These involve procedures which employ various techniques, including externally driven cutters, to remove plaques harder and stiffer than those which can be effectively treated by balloon angioplasty.[3-7]

One risk of any atherosclerosis treatment is damaging healthy arterial tissue. This type of damage may occur as a result of the catheter's presence in the coronary artery and may be instrumental in restenosis.[4] It would therefore be desirable to minimise the invasive time and the size and the flexural stiffness of the catheter device.

The advent of microtechnology invites the possibility of new approaches which could reduce damage caused by the presence of the catheter. The primary disadvantage of using active microdevices (e.g. actuators and pumps) as distinct from externally driven versions, is the relatively low power output achieved by millimetre and, especially, sub-millimetre scale devices. It is evident from published work on existing electrostatic microactuator technologies,[8-11] that the highest reported torques are generated by the harmonic or wobble actuator, although the corresponding operating speeds are low. In addition to the potential to realise useful torques, the wobble motor is mechanically robust by comparison with other electrostatic actuator designs, and the design lends itself to extension of the axial dimensions to increase the output power.

2. DESIGN

The electrostatic harmonic motor utilises the strongest component of the electrostatic force to produce rotational motion. Once the voltage and gap have been optimised, there are two ways of increasing further the output power to volume ratio of the actuator:

1.   maximise the surface area contained within the motor envelope, and
2.   maximise the efficiency.

Like all electrostatic actuators, wobble motor forces are generated by the interaction of surface charge rather than effects produced by volume charge. In the wobble motor, multiple concentric rotor-stator surface pairs can significantly increase the output power without affecting the overall actuator dimensions. This is shown in Fig. 1 for a rotor rolling between two stators, thus utilising both inner and outer rotor surfaces to almost double the force generated.

The efficiency of the actuator can be increased by reducing the system losses. Any friction reducing measure satisfies this criterion. Further gains may also be achieved with a closed loop control system which takes loading or inertial effects on the dynamic performance into account. To date, no reports exist of rotating microactuators operated successfully in closed-loop control systems. Open-loop actuation of a synchronous device relies on the rotor possessing negligible momentum at the end of each actuation step i.e. stepper mode operation.

Closed-loop excitation requires knowledge of the rotor-stator contact point.[12] It has

PLAN VIEW

ROTOR
DIELECTRIC
INNER STATOR
OUTER STATOR

Fig. 1 Twin stator, single rotor wobble motor.

been suggested[13] that it may be possible to monitor the rotor position by sensing the change in rotor-stator capacitance. This method was considered for a hypothetical outer rotor wobble motor, 1 mm tall with a radius of 1 mm. The stator was assumed to be coated with 10 μm thick dielectric, relative permittivity of 4, and a rotor of radius 10 μm larger than the stator was used.

While many techniques may be implemented to measure current changes, these changes will be very small when the variation in capacitance (even for the relatively large hypothetical motor), for the contact moving from the edge of an electrode to the centre is only 1.2 pF. Without electronics close to the actuator to monitor the current, it is unlikely that the change in current due to rotation will be distinguishable from transient currents due to the interconnect cables from the drive electronics and the general noisy environment associated with the unshielded pulses of 100 Volts or more, at frequencies of many kHz, required to drive the motor.

A much simpler method of monitoring position would be to introduce a sequence of contact pads to provide rotor position signals at discrete intervals. Unfortunately, the additional wiring required for such an arrangement is prohibitive for microfabrication.

A variation on the system of contact pads can be introduced into a conventionally designed wobble motor.[14] By segmenting the bearing and connecting segments in two sets such that adjacent segments are in different sets, rotor position information can be provided by the momentary electrical continuity that appears between adjacent segments as the rotor rolls over the inter-segment gap. The segment gaps should be arranged to provide position information at the critical switching instant as shown in Fig. 2. Whilst this method does not provide absolute or continuous position information, it allows the rotor to be tracked from a known starting position, and it is therefore sufficient to allow closed-loop control. With two sets of bearing segments, only one additional cable is required.

The prototype actuator design, incorporating twin concentric rotor-stator surface pairs and a segmented bearing for integrated synchronous control, is shown in Fig. 3, with dimensions (in microns) listed in Table 1. The radial dimensions are as large as possible within the constraints set by the requirement of eventual integration into the catheter system.

Table 1   Prototype dimensions

S	H_s	H_r	R_1	R_2	R_3	R_4	R_5	R_6	Offset	R_7	R_8	R9	R_{10}
40	260	280	865	965	1035	1120	1185	1285	5	975	1025	1125	1175

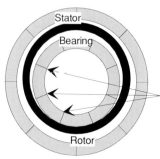

Pulse generated at each of these positions, corresponding to mid-stator electrode. Mid-stator electrode is significant since at this point no torque is applied to the rotor. If the rotor rolls past the centre of the electrode a braking rather than driving torque is applied.

Fig. 2   Critical instant for control.

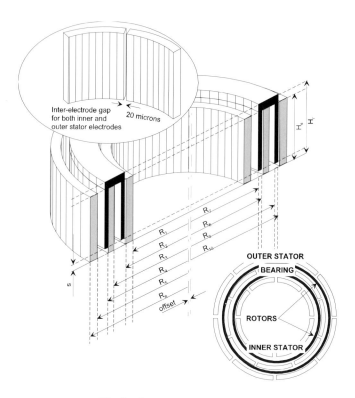

Fig. 3   Prototype wobble motor.

3. FABRICATION

Deep X-ray lithography and optical lithography, combined with nickel electroforming (the LIGA technique[15-19]) has been used for actuator fabrication. Rotors and stators are fabricated separately and assembled manually.

Referring to Fig. 4, the stator is fabricated by optically patterning thin Cr and Ag layers which form the plating base for the microstructures as well as the test connections. A Ti sacrificial layer is then patterned. In addition to producing undercuts in the microstructure, where tracks can be passed from the outer areas to the inner stator and bearing electrodes, the Ti is required for PMMA adhesion. Finally, a thick layer of PMMA is patterned using X-ray lithography and used as a negative to produce the Ni microstructures. The PMMA is then removed.

Rotors, Fig. 5, are fabricated on an unpatterned Ti layer. The Ti allows the rotors to be completely released from the substrate after electroforming. A thick layer of optical resist is patterned and Ni is electroformed into the resulting resist negative. This layer joins the two subsequent cylinders fabricated with X-ray lithography, which form the inner and outer rotor

Altogether, the fabrication process requires two X-ray masks and three optical masks.

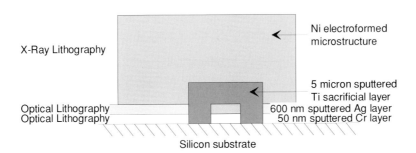

Fig. 4   Stator fabrication section

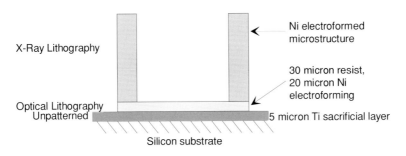

Fig. 5   Rotor fabrication section.

4. ANALYSIS

For rotation, the general expression relating torque, Γ, to energy, U, and angle rotated, θ, is

$$\Gamma = \frac{dU}{d\theta}$$

This formula can be applied to the wobble motor[11] to predict the maximum torques available, however this type of analysis assumes the load is a pure torque. In situations where the translation motion of the rotor also does work, a more general analysis is required. To assess whether or not the translation motion indeed represents a load requires analysis of the system dynamics.

A numerical dynamic model based on Dhuler *et al*,[14] has been developed to combine analytic descriptions of the system mechanics with finite element electric field predictions. Figure 6 shows the static torque function for the fabricated prototype, calculated using 3D finite element analysis.

It is apparent that the overall performance of microactuators is often dominated by friction. Whilst the finite element electric field models can be calculated to <<1% accuracy in 3D, such large models are often wasteful of computing resources when friction can only be defined approximately. Static torque predictions of the prototype wobble motor, differing by less than 3% compared to a 54 000 element model, can be achieved using only 2000 elements.

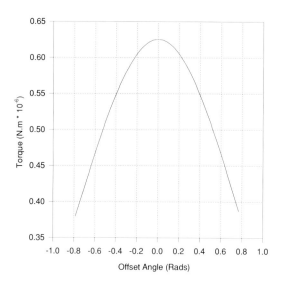

Fig. 6   Prototype torque predictions.

5. RESULTS OF PROTOTYPE FABRICATION

Figure 7 shows SEM photographs of the rotor (inverted) and stator. Figure 8 shows an optical photograph of an assembled actuator.

While the actuators have been made to rotate successfully, their performance is not yet sufficiently repeatable to allow objective measurements to be made. The top surface of the bearing and the inside top surface of the rotor (the two contact surfaces) are unpolished, leading to friction and keying problems which dominate the performance. Speeds of up to 600 rpm have been achieved with open-loop actuation.

At present, the actuators have not been operated in a closed-loop control system. Open-loop operation has demonstrated detectable output position pulses, but sliding friction problems prevent the rotor from rolling smoothly around the perimeter of the bearing.

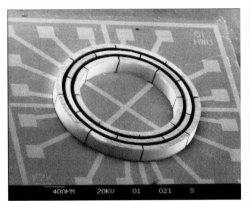 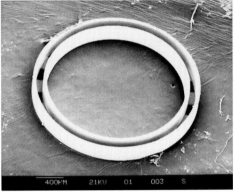

Fig. 7   SEM photographs of prototype stator and rotor.

Fig. 8   Assembled prototype actuator.

6. PERFORMANCE CHARACTERISATION AND DESIGN DEVELOPMENT

A system to address the problem of sliding friction is currently under development. This aims to lift the rotor off the bearing top surface. The same system can apply precise loads to the rotor for measuring output torques and forces.

7. SUMMARY

A high power electrostatic rotating micro actuator, with a mechanism for integrated synchronous control, has been designed and fabricated with a view to integration in a catheter system for removing arterial plaque from the coronary arteries. Theoretical finite element analysis of the electric fields and gross generated torques suggest useful outputs close to 1 mNm could be realised in future. Prototype actuators rotate successfully, however steps to reduce sliding friction should significantly improve the performance.

REFERENCES

1.   M.J. Davies and N. Woolf: *Atheroma: Atherosclerosis in Ischaemic Heart Disease. Volume 1: The Mechanisms*, London: Science Press Limited, 1990.
2.   R.M. Nerem: 'Vascular Fluid Mechanics, the Arterial Wall, and Atherosclerosis', *J. Biomechanical Engineering*, Transactions of the ASME, 1992, **114,** 274–282.
3.   E.B. Carlson: 'Coronary Angioplasty: An alternative to Bypass Surgery', *Proceeding of the IEEE*, 1988, **76**, 1194–1203.
4.   T.A. Fischell and M.L. Stadius: 'New Technologies for the Treatment of Obstructive Arterial Disease', *Catheterization and Cardiovascular Diagnosis*, 1991, **22**, 205– 233.
5.   P.J. de Feyter, D.C. MacLeod, D. Foley, P.P.T. de Jaegere and P.W. Serruys: 'Interventional Techniques for the Management of Coronary Artery Lesions: An Update', *Clinical Cardiology*, 1993, **16**, 586–593.
6.   D.C. MacLeod, M. de Jong, V.A. Umans, J. Escaned, R.J. Suylen, P.J. de Feyter: 'Directional Athererctomy: Combining Basic Research and Intervention', *American Heart Journal*, 1993, **6**, 1748–1759.
7.   P.S. Gilmore, T.A. Bass, D.A. Conetta, R.F. Percy, Y.G. Chami, B.J. Kircher and A.B. Miller: 'Single Site Experience with High-Speed Coronary Rotational Atherectomy', *Clinical Cardiology*, 1993, **16**, 311–316.
8.   T. Furuhata, T. Hirano, L.H. Lane, R.E. Fontana, L.S. Fan and H. Fujita: 'Outer Rotor Surface-micromachined Wobble Micromotor', *Proceedings of the 1993 IEEE Micro Electro Mechanical Systems – MEMS*, Fort Lauderdale, 1993, 161–166.
9.   M. Mehregany and Y.C. Tai: 'Surface Micromachined Mechanisms and Micromotors', *Journal of Micromechanics and Microengineering*, 1991, **1**, 73–85.
10. M. Mehregany, S.F. Bart, L.S. Tavrow, J.H. Lang and S.D. Senturia: 'Principles in Design and Microfabrication of Variable-capacitance Side-drive Motors', *Journal of Vaccuum Science Technology*, 1990, **A8**, 3614–3624.

11. W. Trimmer and R. Jebens: 'Harmonic Electrostatic Motors', *Sensors and Actuators*, 1989, **20**, 17–24.

12. S.F. Bart and J.H. Lang: 'Analysis Of Electroquasistatic Induction Micromotors', Sensors and Actuators, 1989, **20**, 97–106.

13. S.C. Jacobsen, R.H. Price, J.E. Wood, T.H. Rytting and M. Rafaelof: 'Design Over view Of An Eccentric-Motion Electrostatic Microactuator (The Wobble Motor)', *Sensors and Actuators*, 1989, **20**(1–2), 1–16.

14. V.R. Dhuler, M. Mehregany and S.M. Phillips, 'Experimental Technique and a Model for Studying The Operation Of Harmonic Side-Drive Micromotors', *IEEE Transactions on Electron Devices*, 1993, **40**(11), 1977–1993.

15. J. Mohr, C. Burbaum, P. Bley, W. Menz and U. Wallrabe: 'Movable Microstructures Manufactured by the LIGA Process as Basic Elements for Microsystems', *1st Conference on. Micro, Electro, Opto, Mechanic Systems and Components*, Berlin, 1990, 529–537.

16. A. Rogner, J. Eicher and R.P. Peters: 'The LIGA Technique - What Are The New Opportunities', *MME '92 - Third European Workshop on Micromachining, Micromechanics and Microsystems*, Leuven, Belgium, 1992, 118–140.

17. W. Ehrfeld and D. Münchmeyer: 'LIGA Method', *Nuclear Instruments and Methods in Physics Research*, 1991, **A 303**, 523–531.

18. J. Mohr, P. Bley, M. Strohrmann and U. Wallrabe: 'Microactuators Fabricated by the LIGA Process', *Proceedings of: Actuator 92 3rd International Conference on New Actuators*, Bremen, 1992, 19–23.

19 U. Wallrabe, P. Bley, B. Krever, W. Menz and J. Mohr: 'Theoretical and Experimental Results of an Electrostatic Micromotor with large Gear Ratio Fabricated by the Liga Process', *Micro Electro Mechanical Systems '92*, Travemunde, Germany, 1992, 139–140.

ACTIVE VIBRATION CONTROL USING THIN ZnO FILMS

D.F.L. JENKINS*, M.J. CUNNINGHAM, W.W. CLEGG*,
G. VELU† AND T.D. REMIENS†

*Division of Electrical Engineering, Manchester School of Engineering,
University of Manchester Oxford Road, Manchester M13 9PL*
*Centre for Research in Information Storage Technology, University of Plymouth,
Drake Circus, Plymouth PL4 8AA*
†Université de Valenciennes, CR1TT Maubeuge, Laboratoire des Matériaux,
Advancés Céramique, ZI Champ de l'Abesse, 59600 Maubeuge, France*

ABSTRACT

An active vibration control system has been developed based upon analogue feedback and optical vibration sensing. ZnO thin films have been deposited using r.f. magnetron sputtering onto silicon substrates and cantilevers fabricated using photolithography. These cantilevers provide a means of actuation and micro-positioning. The system operates such that unwanted vibrations in the cantilever are removed and yet it remains possible to deflect the cantilever statically or dynamically as required. Results are presented for such a system.

1. INTRODUCTION

Digital active vibration control techniques have previously been applied to large structures, where the frequency of vibration is low.[1] In such structures, for example cantilevers, up to three ceramic PZT elements may be incorporated. One of these may be used for sensing and one or more used for actuation. One of the problems associated with this is the optimum location of sensors and actuators. This is eliminated by the use of a single piezoelectric actuating element. This is possible by using optical sensing and electronics to combine the control and positioning signals. As structures become increasingly small in size their natural frequencies of vibration become much higher and analogue control becomes more attractive than digital methods.[2]

For micro-structures ZnO films are extremely promising as micro-mechanical actuators and there are many potential application areas such as scanning mirror drives,[3] micro surgery[4, 5] and scanning force microscopy.[6] In scanning force microscopy, and in particular magnetic force microscopy, it necessary to implement simultaneous active vibration control and micro-positioning.

To investigate the feasibility of applying such an active vibration control system to much smaller structures, silicon cantilevers incorporating ZnO thin films were fabricated.

2. ACTIVE VIBRATION CONTROL CONSIDERATIONS

A number of feedback strategies have been proposed for active vibration control. These range from simple techniques such as constant gain or variable gain feedback[7] to control algorithms using neural networks[8] and fuzzy logic.[9] In this work constant gain feedback was used to control the cantilevers first two bending modes. Attempting to control the higher order modes can degrade control performance.[10] For a magnetic force microscope the typical natural frequency of vibration of the cantilever is 66 kHz. The second mode of vibration is therefore at about 400 kHz.[11] Current digital signal processors are not fast enough to implement suitable control algorithms to suppress both of these modes, and so a suitable control system has been developed using analogue electronics. To monitor the cantilever end displacement, the optical lever effect was used.[12] Light from a diode laser at 670 nm was focused onto the mirror and the reflected light collected by a position sensing photodetector. For small displacements, such as occur in this application, the response of the position sensing detector can be assumed to be linear. The sensor signal must then filtered, amplified and phase shifted to provide the feedback signal. The sensor signal was filtered using high and low pass active filters. These were 4th order Bessel filters which have the best phase linearity. The feedback signal is phase shifted by 180° using a (switchable) inverting amplifier. Figure 1 shows typical open and closed loop response obtained using this configuration. The rapid attenuation of the vibration can be clearly seen under closed loop conditions. The optical beam deflection sensing system was calibrated using a PZT bimorph element. The end displacement, Δx, of a such a series bimorph element may be determined from its physical properties and the exciting voltage, and is given by:[13]

$$\Delta x = \frac{3d_{31}L^2}{2t^2}V$$

where d_{31} is the piezoelectric charge constant (m/V), L is the length of the cantilever (m), t is the bimorph thickness (m) and V is the applied voltage (V).

3. ACTUATOR FABRICATION

ZnO films were deposited onto oxidised (100) silicon wafers upon which a bottom electrode (Ti/Pt) had been previosly deposited using r.f. magnetron sputtering.[14] The ZnO films were deposited, also by r.f. magnetron sputtering, at 500 °C at a rate of 50 Åmin[-1].

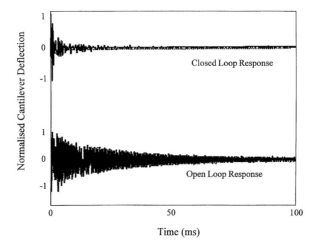

Fig. 1   Typical open and closed loop impulse responses.

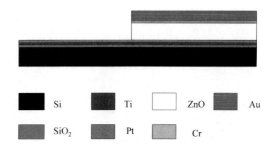

Fig. 2   Schematic of the Silicon–ZnO cantilever, not to scale.

The films are strongly c-axis orientated and colunar with columns perpendicular to the substrate.[15] The substrates was etched to enable the top electrode (Cr/Au) to be deposited. The top electrode was patterned using photolithography to enable the final cantilevers to be formed. Figure 2 shows the cross-section of a typical cantilever (not to scale).

4. ACTUATION AND VIBRATION CONTROL

The experimental arrangement used in this work is depicted schematically in Fig. 3. When a positioning signal is applied to the piezoelectric film the cantilever is forced into periodic motion at the drive frequency. The strain induced in the cantilever is dependent on the magnitude and frequency of the applied voltage. A 5 V amplitude sinusoidal positioning signal (Feedback TWG300) caused the end of the cantilever to move with an amplitude of 250 nm at a frequency of 10 Hz. The amplified signal from the photodetector

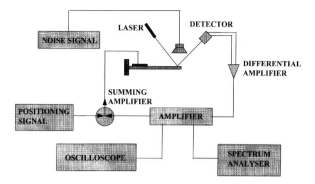

Fig. 3   Experimental arrangement.

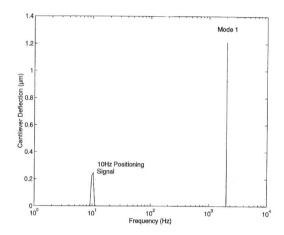

Fig. 4   Open loop modal response.

was sampled by the PC3OPG A/D converter and its fast Fourier transform (FFT) subsequently obtained in order to determine the frequency components. The fundamental frequency of vibration of this cantilever was 2 kHz. To simulate the effect of operating this cantilever in an acoustically noisy environment the cantilever was also disturbed by a loudspeaker operating at the cantilever resonance frequency Fig. 4 shows the open loop spectrum obtained for the cantilever. The actuation signal at 10 Hz and the first mode at 2 kHz are clearly seen. Active vibration control was applied to the cantilever by feeding back a suitably amplified feedback signal to the piezoelectric element. The amplifier frequency response was tailored in such a way that the feedback was inoperative at the positioning signal frequency and also at frequencies beyond that of the second bending mode. The effect of active vibration control can be seen from the closed loop spectrum in Fig. 5. This figure shows that the cantilever can be positioned as required by the positioning signal, without change of amplitude but that the vibration signal is constantly canceled.

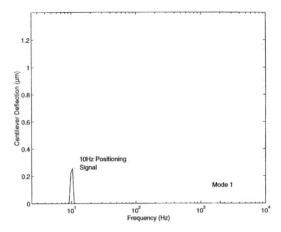

Fig. 5   Closed loop modal response.

5. DISCUSSION AND CONCLUSIONS

It has been possible to use composite ZnO piezoelectric thin films for simultaneous active vibration control and micro-positioning of silicon micro-structures. The current arrangement enables the end of the cantilever to be deflected by up to 700 nm, whilst active vibration control is implemented. The amount of deflection is dependent on the cantilevers properties as well as those of the piezoelectric film.

ACKNOWLEDGEMENTS

The authors acknowledge UK EPSRC for financial support and Keith Birtwistle (Manchester School of Engineering) for cantilever fabrication.

REFERENCES

1. V.K. Varadan, S.-Y. Hong and V.V. Varadan: 'Piezoelectric sensors and actuators for active vibration damping and digital control', *IEEE, 1990 Ultrasonics Symposium*, 1990, 1211–1214.
2. D.F.L. Jenkins, M.J. Cunningham and W.W. Clegg: 'Analogue vibration cancellation by means of piezoelectric elements', *Journal of Structural Control*, 1996, **3**(1–2), 45–51.
3. T. Tominanga, N. Ohya, K. Senda, T. Idogaki and T. Hattori: 'Flexible stacked type actuators', *5th International Symposium on Micro Machine and Human Science Proceeding's (IEEE)*, 1994, 143–7.
4 H. Miyazaki, T. Kemeya, T. Sato, Y. Hatamura and H. Morishita: 'Construction of

an ultra-micro-manipulation system based on visual control – realisation of nano-hand-eye system', *1994 IEEE Symposium on Emerging Technologies and Factory Automation*, 1994, 74–7.

5. W. Menz: 'Three-dimensional microstructures in various materials for medical applications', *1993 International Con frrence on Systems, Maim amml Cybernetics*, 1993, 417–22.

6. M.J. Cunningham, S.T. Cheng and W.W. Clegg: 'A differential interferometer for scanning force microscopy', *Meas. Sci. and Technol.*, 1995, **5**(1), 350–54.

7. E..F Crawley and J. de Luis: 'Use of piezoelectric actuators as elements of intelligent structures', *AIAA Journal.*, 1987, **25**(10), 1373–1384.

8. V. Rao, R. Damle, C. and F. Kern: 'The adaptive control of smart structures using neural networks', *Smart Materials and Structures*, 1994, 3(3), 354–366.

9. I. Zeinoun and F. Khorrami: 'An adaptive control scheme based on fuzzy logic and its application to smart structures', *Smart Materials and Structures*, 1994, **3**(3), 266–276.

10. A. Preumont: 'Spillover alleviation for non-linear active control of vibration', *J. Guidance*, 1988, 11(2), 124–130.

11. G.Y. Chen, R.J. Waarmack, T. Thundat, D.P. Allison and A. Huang: 'Resonance response of scanning force microscopy cantilevers', *Rev. Sci. Instrum.*,1994, **65**(8), 2532–37.

12. D.F.L. Jenkins, M.J. Cunningham, W.W. Clegg and M.M. Bakush: 'Measurement of the modal shapes of inhomogeneous cantilevers usine optical beam deflection', *Meas. Sci. and Technol.*, 1995, **6**, 160–166.

13. Morgan-Matroc Ltd., Thornhill, Southampton, S09 5QF, UK. Piezoelectric Ceramics – Technical Literature.

14. D. Remiens, J.F. Tirlet, B. Jaber, B. Thierry and C. Moriainez: 'Single target sputter deposition and post-processing of perovskite lead titanite thin films', *J. Eur. Ceram. Soc.*, 1994, **13**, 493–500.

15. D.F.L. Jenkins, M.J. Cunningham, W.W. Clegg, G. Velu and D. Remiens: 'The use of sputtered ZnO piezoelectric thin films as broadband micro-actuators, *Sensors and Actuators A*, 1997, **63**, 135–139.

THE PRINCIPLE AND OPERATION OF A RESONANT SILICON GYROSCOPE

D. WOOD*, G. COOPER*, J.S. BURDESS†, A.J. HARRIS§
AND J.L. CRUICKSHANK§

*School of Engineering, University of Durham, South Road, Durham, DH1 3LE
†Department of Mechanical Engineering, University of Newcastle-upon-Tyne,
Stephenson Building, Newcastle-upon-Tyne, NE1 8ST
§Department of Electrical and Electronic Engineering, University of
Newcastle-upon-Tyne, Merz Court, Newcastle-upon-Tyne, NE1 7RU

ABSTRACT

In this paper a new type of silicon resonant gyroscope is presented that is simple to both manufacture and operate. The structure of the gyroscope is described, and its method of operation explained. Details on the fabrication process developed to make the test samples are given, and the structure has demonstrated a rate of turn measurement capability.

INTRODUCTION

The silicon resonant gyroscope is one of the most exciting developments in microengineered sensors for many years. The device combines principles of mechanical resonance, Coriolis force, electronic phase locked loops and the symmetry of crystalline silicon to produce a means of measuring angular velocity in a device measuring only a few millimetres square.

Great demand for these devices is expected from the automotive sector, where an inexpensive means of measuring a rate of turn about the three principle axes of a car would provide important feedback to automatic control of the vehicle's steering, suspension and braking systems. Further demand is expected from the military sector for smart munitions, the virtual reality sector for measuring orientation of data gloves, and the petrochemical exploration sector to give feedback on the attitude and positioning of drill tips. No doubt there will be many more applications which require a means of measuring rate of turn within a compact, inexpensive package.

Silicon gyroscopes are devices with two degrees of freedom, and hence are much more difficult to design, manufacture and test than many other microengineered structures, such as accelerometers. This has limited the successful demonstration of a device to just a few laboratories, with gyroscope implemented in structures such as a tuning

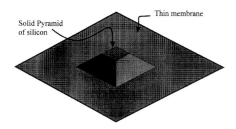

Fig. 1   A membrane gyroscope.

fork,[1] a vibrating beam[2] and a ring.[3] Presented here is a structure based on a vibrating silicon membrane: only three masks are required to make this device, making it, in manufacturing terms, one of the simplest gyroscopes reported to date.

DESIGN DESCRIPTION

The basic structure of the gyroscope is shown in Fig. 1. A pyramid shaped mass of solid silicon is supported in the centre of a thin square silicon membrane. The outer edge of the membrane remains attached to the silicon wafer (not shown).

The flexibility of the membrane allows the pyramid some freedom to rock from side to side. By applying equal and opposite forces on either side of the membrane, in a direction normal to the membrane, it is possible to excite this rocking motion. This situation is shown in Fig. 2, where the primary resonance is shown as a rocking motion about the x-axis. As a result of the rocking motion the centre of gravity of the pyramid is made to move back and forth along the y-axis.

If the whole structure is then made to rotate about the z-direction (Fig. 2b), a Coriolis force will be generated. The direction of the Coriolis force will be orthogonal to the rate of turn and to the direction of motion of the centre of gravity of the pyramid, and will therefore act along the x-direction. The Coriolis force will therefore cause the structure to begin resonating with the same rocking mode, but about the y-axis. This situation is shown in Fig. 2. The motion of the device is not dissimilar to the original Foucault pendulum, with the pyramid being the equivalent of the pendulum bob.

Due to the symmetry of the design the primary and secondary modes of the structure are interchangeable. This automatically means that in the ideal case they will have exactly matched resonant frequencies, the optimum condition to ensure a maximum transfer of energy from the primary to secondary modes.

DRIVE AND SENSE ARRANGEMENT

Driving the gyroscope into resonance is achieved via a pair of electrodes positioned underneath the membrane on either side of the pyramid. These electrodes are close to,

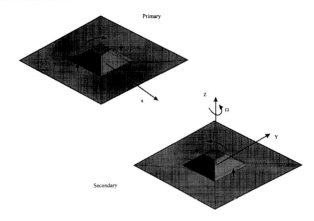

Fig. 2   The primary and secondary resonances of the gyroscope.

but not in contact with, the membrane. The membrane itself is also metallised and acts as a third electrode. With this arrangement it is possible to electrostatically drive the membrane into resonance by applying sinusoidal voltages to the electrodes (Fig. 3).

A similar pair of electrodes positioned on the other two sides of the membrane act as capacitive displacement sensors to measure the amplitude of the secondary resonance, giving the measure of rate of turn. In order to check that the structure would move in the manner described a finite element (FE) model of the device was constructed.

FABRICATION

The fabrication of this device involves three stages. In the first stage the resonant element is manufactured. The electrodes are fabricated on a separate wafer, and finally the two components are assembled.

The gyroscope resonant structure is fabricated from (100)-oriented silicon. The first stage of the fabrication process is to clean the wafer, and then oxidise it to a thickness of about 0.8 μm on both sides. The thickness of this oxide layer is not critical as it serves only as an etch mask and does not form part of the final structure. The next stage of the process is to remove the silicon dioxide from the mirrored surface of the wafer using HF. A high concentration of boron dopant is then diffused into the front surface of the wafer, using a solid source of boron. During this process the silicon dioxide on the reverse side of the wafer acts as a diffusion barrier, preventing the boron from entering the back of the wafer. A window is then opened in the oxide layer on the back side of the wafer. The purpose of this window is to expose selected areas of silicon on the back side of the wafer, allowing the silicon to be removed by the anisotropic etchant. The next stage of the process is to metallise the boron diffused side of the wafer. This metal layer acts as the top electrode in the capacitive drive/sense arrangement shown in Fig. 3. The metal layer is formed by evaporating first a layer of chromium and then gold.

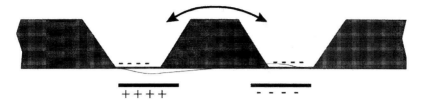

Fig. 3   The electrostatic excitation/detection arrangement.

This combination is used as they are both resistant to the subsequent anisotropic etch process, performed using an EDP etch solution. The etch proceeds vertically through the wafer until it reaches the boron diffused region on the front side. Once the etch front reaches a region of sufficiently high boron concentration the etch rate dramatically reduces, and the etch effectively stops. In the horizontal direction the etch is controlled by the inclined {111} planes which run through the wafer at 54.7 degrees to the surface,and attacked at a much slower rate than the horizontal (100) plane. The resulting structure is shown in Fig. 4. This process typically takes around 5 hours to produce a boron doped membrane free of any residual silicon. The point at which the etch front is stopped by the boron diffusion controls the thickness of the silicon membrane. With a 4 hour diffusion the resultant membrane thickness is 2.3 μm.

Figure 5 shows the mask design used to define the membrane and the central mass. The truncated pyramid mass is bounded by four {111} planes meeting at convex corners, and the membrane is bounded by the same four planes meeting at concave corners. An unwanted feature of the etch process is that the convex corners are attacked at a significant rate. This has the effect of reducing the size of the pyramid and making it non-symmetrical. In order to correct for this problem the corners were built out as shown: these areas are etched away during the process but they remain in place long enough to protect the corners of the pyramid.

In the mask shown all of the areas coloured black are etched away. The black lines around the edge of the structure are etched into V grooved trenches and allow the central square containing the gyroscope to be separated from the silicon wafer. In addition to the corner compensation Fig. 5 shows four diagonal beams running from the corners of the pyramid across the membrane. These beams also contribute to the corner compensation. However their main purpose is to provide mechanical support to the structure during the assembly process.

As the beams are not aligned with the crystal planes the silicon beneath them is dissolved away, leaving a free standing oxide structure. These beams are left in place during the assembly process, and help to prevent the membrane from fracturing while the wafer is diced. Once the membrane is assembled with the drive electrodes the whole structure becomes more robust and the beams may be removed. This is done by touching the beams with a pin to fracture them and then blowing the broken beam away with a nitrogen gun.

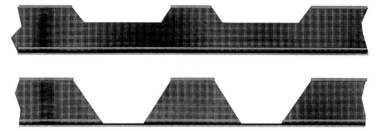

Fig. 4   The etch partially and fully completed.

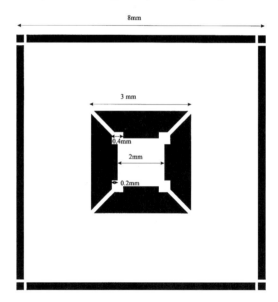

Fig. 5   The mask pattern used to etch the membrane. The dimensions
shown correspond to those of the final structure.

The second component of the structure is the electrode arrangement containing the
drive and sense electrodes. This is fabricated on a Pyrex glass disc. After a cleaning
stage the disc is metallised with a chromium/gold combination. This metal layer is used
as an etch mask in order to form the spacing trench in the surface of the disc. A more
conventional technique using a simple photoresist mask was found to have insufficient
adhesion to survive the long etch time used to make the trench. The gold and chromium
are patterned using separate wet etching processes. The next stage is to etch the trench in
buffered HF. A trench depth of about 1.8 μm is produced after a 1 hour etch time. Following the trench etch the remaining metal is removed, leaving a bare disc with a trench
patterned in the surface.

The pattern of the trench is shown in Fig. 6. The central square region sits immediately below the membrane while the arms extending from this region act as channels
along which the tracks connecting to the capacitor plates are routed.

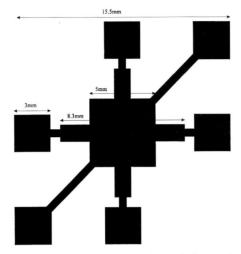

Fig. 6   The mask pattern used to etch the trench.

The etched Pyrex disc is then metallised again with chromium/gold and patterned into the desired arrangement for the electrodes. This arrangement has a total of 8 connections as shown in Fig. 7. Four of these go directly to the drive/sense plates and another two are connected to a guard ring running between all of the electrodes. All of these are positioned in the bottom of the etched trench. The two other connections are not in the trench, but instead sit on the wafer surface. This allows an electrical contact to be made with the metal on the membrane. The guard ring was found to be necessary in order to minimise electrical coupling between the 4 plates.

The final stage is in assembling the membrane and electrodes. For the test structures this is done by eye and the membranes were fixed in place with a drop of adhesive at each corner. Figure 8 shows a cross section of the completed device.

Figure 9 shows the electrode arrangement, light grey, superimposed on the trench pattern in dark grey. Note that the electrodes marked top plate contact are not lying in a trench.

TESTING

The completed device was tested optically, electrically and finally on a turntable. It was necessary to hold the device under vacuum in order to prevent air from damping the resonance. The need to operate the device in vacuum is common to most resonant gyroscopes. Details of the testing procedures, results and the control electronics have been published in more detail elsewhere.[4, 5] The device was non-linear, but despite this and a non-perfect matching in the resonant modes, it was possible to operate the device as a gyroscope. The output voltage as a function of rotation rate is shown in Fig. 10.

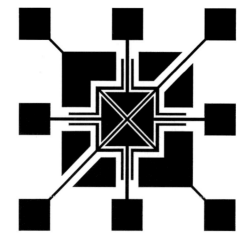

Fig. 7 The electrode arrangement.

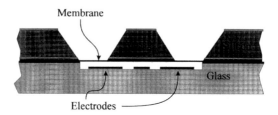

Fig. 8 A cross-section of the completed device.

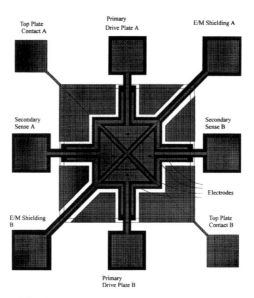

Fig. 9 A plan view of the completed structure.

**Voltage Output Plotted Against Rotation Rate When
Primary Amplitude = 270mV**

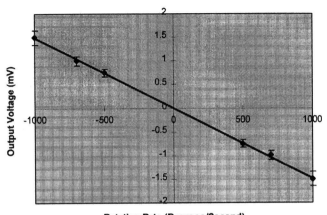

Rotation Rate (Degrees/Second)

Fig. 10 The results of the rotation rate tests.

CONCLUSION

A new design of silicon gyroscope based on a capacitively driven and sensed membrane has been presented. Fabrication details have been described, and the whole procedure uses standard processes with relaxed geometrical constraints on the lithography, and only three masks – one for the silicon and two for the pyrex disc. The design described has been manufactured, tested and shown to work as a gyroscope. Ways to improve on the performance of the structure are being investigated with further work.

ACKNOWLEDGEMENTS

This work was performed with a grant provided by EPSRC and the Defence Research Agency at Farnborough. The authors are grateful to both these bodies for their support.

REFERENCES

1.   J. Bernstein, S. Cho, A.T. King, A. Kourepenis, P. Maciel and M. Weinberg: 'A Micromachined Comb-Drive Tuning Fork Rate Gyroscope', *Micro Electro Mechanical Systems Workshop*, Technical Digest #93CH3265-6, IEEE, February, 1993, 143–148.
2.   K. Maenaka and T. Shiozawa: 'A Study of Angular Rate Sensors Using Anisotropic Etching Technology', *Sensors and Actuators*, 1994, **45**, 72–77.

3.   M.W. Putty and K. Najafi: 'A Micromachined Vibrating Ring Gyroscope', *Solid State Sensor and Actuator Workshop*, IEEE, 1994, **6**, 213–220.
4.   A.J. Harris, J.S. Burdess, D. Wood, J.L. Cruickshank and G. Cooper: 'A Vibrating Silicon Diaphragm Micromechanical Gyroscope', *Elec. Lett.*, 1995, **31,** 1567–1568.
5.   D. Wood, G. Cooper, J.S. Burdess, A.J. Harris and J.L. Cruickshank: 'A Silicon Membrane Gyroscope with Electrostatic Actuation and Sensing', *SPIE Conference Micromachining and Microfabrication '95*, Austin, Texas, 23-25 October, 1995, vol. 2642, 74–83.

P–N AND SCHOTTKY JUNCTION CANTILEVER ACTUATORS

S. BETTERIDGE, T.J. LEWIS, J.P. LLEWELLYN
AND M.J. VAN DER SLUIJS

Institute of Molecular and Biomolecular Electronics, University of Wales Bangor,
Dean Street, Bangor, Gwynedd, LL57 1UT.

1. INTRODUCTION

Over the last decade much effort has gone into forming piezoelectric materials, eg PZT, on top of micromachined silicon to make sensors and actuators. However the techniques involved in making such devices are complicated and time consuming, reducing their commercial value. This paper demonstrates that a separate piezoelectric material is not required to produce a silicon actuator compatible with contemporary electronics.

The Lippmann equation (eqn (1)), which is derived from the Gibbs adsorption equation at constant temperature, pressure and composition, predicts that in all interfaces a change in interfacial tension ($\Delta\gamma$) is generated when there is a change in the applied potential difference (ΔV) across it.

$$\Delta\gamma = -q\Delta V \tag{1}$$

where $\pm q$ is the charge separation across the interface. This equation therefore implies that all interfaces have piezoelectric qualities.[1]

Large area silicon p–n and Schottky junctions have been chosen to demonstrate this effect, because they are considered to have well defined junction interfaces. The junctions were mounted to form cantilevers, as shown in Fig. 1. An alternating potential (V_{AC}) and bias voltage (V_B) applied across the junction interface would, according to eqn (1), cause an alternating tension and thus a strain in the plane of the junction, so causing the cantilever to vibrate.

2. EXPERIMENTAL

The p–n and Schottky junctions used in these experiments were prepared at the Southampton University Microelectronics Centre.[2] The metal layer in the Schottky junction

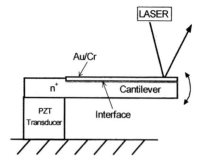

Fig. 1   Schematic diagram of the Schottky junction cantilever and experimental arrangement.

was Cr/Au and both types of junctions had similar dimensions of approximately 25 x 5 x 0.5 mm.

Two techniques were used to measure the magnitude of the induced vibration. The first utilised an optical interferometer to measure the vibration amplitude at a selection of points along the length of the cantilever. The laser beam of the interferometer was reflected from a mirror attached to the top surface (see Fig. 1). The second technique measured the shear force exerted by the vibrating cantilever at its fulcrum using a PZT transducer. In all experiments an alternating voltage of $V_{AC} = 0.34$ V rms was applied across the junctions. The signals produced by the interferometer or the PZT transducer were measured using a lock-in amplifier to minimise background noise.

3. FREQUENCY RESPONSE

Initially, experiments were performed with a bias voltage of 2 V over a range of frequencies to determine the frequency response of the cantilever. The results using the PZT transducer detector for both types of junction are shown in Fig. 2. Resonances can clearly be seen at 980 and 820 Hz which correspond well to the theoretically determined values for the fundamental frequency (f_0) of 986 and 695 Hz for the p–n and Schottky cantilevers respectively. These values of f_0 were derived from the well known expression for cantilever resonance given in eqn (2), where E is Young's modulus and ρ is the density for silicon, and t and l are the thickness and length of the cantilever respectively.

$$f_0 = 0.1615\left(\frac{Et^2}{\rho l^4}\right)^{1/2}$$

(2)

Interferometer measurements of the amplitude of the vibration of the p–n junction cantilever as a function of both frequency and position along the cantilever are shown in Fig. 3. As expected the maximum amplitude is at the free end and decreases to a minimum, rather than zero, at the fixed end.

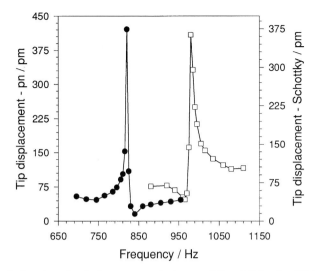

Fig. 2   Induced mechanical resonances in the p–n (▢) and Schottky (●) junction
cantilevers measured with the PZT transducer.

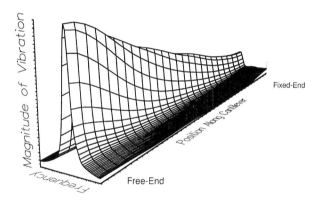

Fig. 3   Induced vibration in the p–n junction cantilever measured
with the optical interferometer.

Second, third and fourth harmonic resonances were also recorded at frequencies up to
32 kHz. There is also good agreement between these harmonic frequencies and those
predicted from theory.[3]

4. BIAS RESPONSE

The amplitude of the vibration at the second harmonic frequency was measured with the
PZT transducer when a range of bias voltages (V_B) were applied to the junction and these
results are shown in Fig. 4. The second harmonic was chosen since the shear force at this

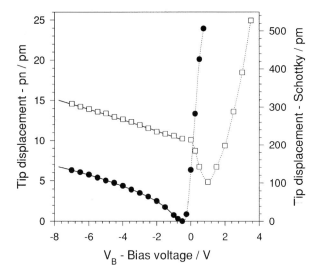

Fig. 4   Effect of bias voltages on the displacement at the end of the p–n (□) and Schottky (●) junction cantilevers. Measurements at the second harmonic frequency. The solid line indicates the fit to eqn (4).

resonance is greater then at the fundamental, even though the actual tip displacement is smaller. The shear force was then converted to an actual tip displacement to give similar results to those obtained from the interferometer.[4] The results for both junctions follow similar trends, with a gradual increase in the tip displacement for reverse bias and a sharp increase for forward bias, which can be related to the general i–v characteristic of a silicon junction.

The reverse bias data was fitted to a theoretical model based upon the Lippmann equation and the well-established capacitance of a silicon junction in reverse bias. This capacitance, see eqn (3), involves both the diffusion voltage V_D and the bias voltage V_B and an index α, which is 0.5 for an abrupt junction and 0.33 for a linearly graded junction.

$$C \propto \frac{1}{\left(V_D + V_B\right)^{\alpha}} \tag{3}$$

By incorporating the equation above with eqn (1) and integrating, an expression for the alternating component of the interfacial tension using the condition $V_{AC} < V_B$, can be obtained, which is given in eqn (4).

$$\Delta\gamma \propto (2 - \alpha)\left(V_D + V_B\right)^{(1-\alpha)} \tag{4}$$

This equation was used to obtain the theoretical fit of the reverse bias data shown in Fig. 4, from which values of α for both types of junction were obtained. The values of 0.33 and 0.45 for the p–n and Schottky junctions respectively, suggest that the doping gradient in the Schottky junction is abrupt, while that of that p–n junction is linear. This agrees with what is expected from the manufacture of the two types of silicon junctions.

It is interesting to note, that the apparent exponential rise in the tip displacement for forward bias in Fig. 4 is expected, because the capacitance of these junctions is also proportional to the exponential of the diffusion and bias voltages in this region. A theoretical fit was not possible however, probably because of the small forward bias voltages used in the experiment.

5. DISCUSSION AND CONCLUSIONS

The experiments clearly show that silicon junctions are piezoelectric actuators, without the need for any additive piezoelectric material. In addition the good fit of the reverse bias data in Fig. 4 indicates that it is the Lippmann effect which is the main mechanism causing the cantilever to vibrate. However other effects such as heating may modify the vibration and a fuller investigation is needed to determine the significance of such effects to the vibration of the cantilever.

The cantilever used in these experiments was quite large in comparison to microchip dimensions, but considering that the techniques for micromachining silicon are well established, it should be possible to scale the cantilever down for applications in micromechanics.

This effect also has a bearing on the long term ageing properties of silicon junctions as repetitive strains at the interface in normal operation may cause defects in the crystalline structure, which may eventually destroy the junction.

ACKNOWLEDGEMENT

This work formed part of an ESPRC ROPA award, No. GRK 62644 receipt of which is gratefully acknowledged.

REFERENCES

1. T.J. Lewis, J.P. Llewellyn and M.J. van der Sluijs: 'Electrokinetic Properties of Metal-Dielectric Interfaces', *IEE Proceedings A*, 1993, **140**(5), 385–392.
2. Funded through SERC pump-priming project PB116.
3. A. Wood: Acoustics, Blackie and Son Ltd, London and Glasgow, 1950.
4. F.H. Newman and V.H.L. Searle: The General Properties of Matter, Edward Arnold and Co., London, 1950.

A LINEARISED ELECTROSTATIC ACTUATOR

M.M. BAKUSH, D.F.L. JENKINS*, M.J. CUNNINGHAM,
C. FERRARI† AND G.P. SCHIONA†

*Division of Electrical Engineering, Manchester School of Engineering,
University of Manchester, Oxford Road, Manchester M13 9PL*
**Centre for Research in Information Storage Technology, University of Plymouth,
Drake Circus, Plymouth PL4 8AA*
†Electrical Engineering Department, University of Ferrara, Italy

ABSTRACT

There is currently considerable interest in positioning miniature structures. For such structures electrostatic actuation is attractive. An electrostatic actuator is however inherently non-linear. The behaviour of such an actuator is theoretically modelled and experimentally verified. A control system has been developed which enables linear electrostatic actuation to be realised.

1. INTRODUCTION

In recent years, there has been a growing interest in actuation that is capable of being used at the micrometer level. Recently, Wood *et al.*[1] reviewed a variety of actuation mechanisms used in microengineering and have shown that electrostatic actuators have a strong position in microengineering applications due to their compatibility with CMOS processes, low temperature dependence and low power consumption. There are however disadvantages, namely:

- The electrostatic force is non linear.
- It is restricted mainly by the dielectric strength of the insulating barrier and is sensitive to the permittivity of the gap separation.
- If air is used as the separating dielectric, variations due to temperature and humidity will occur.

2. THEORETICAL CONSIDERATION

It is well known[2,3] that the relationship between the dc electrostatic displacement and the applied electric field is non-linear. Figure 1 shows the static deflection of a cantilever as a function of the applied d.c voltage. For time varying actuation, it is possible to vary the magnitude of both the bias field and the changing field. Figure 2 shows the effect of

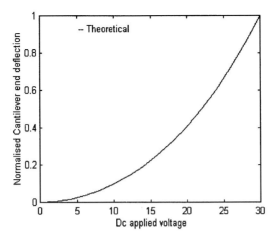

Fig. 1   Static deflection as a function of applied dc bias voltage.

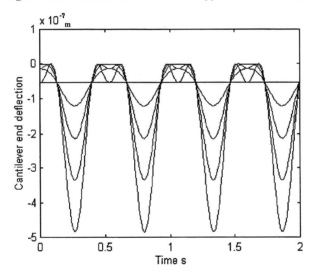

Fig. 2   The predicted cantilever vibration as function of the a.c voltage (0, 1, 4, 5, 7 V)

varying the changing field for a given dc bias field. We have confirmed experimentally[4] that if the square root of the total drive field is applied then the cantilever response is linearised as predicted from theoretical considerations. This is shown in Fig. 3.

3. EXPERIMENTAL INVESTIGATION

The experimental set-up used in this work is shown in Fig. 4. A function generator with

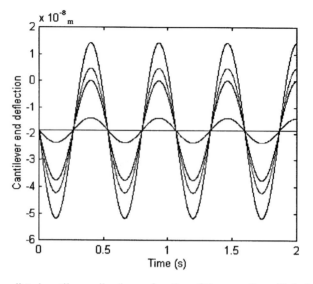

Fig. 3   The predicted cantilever vibration as function of the a.c voltage (0, 1, 4, 5, 7 V) with Vdc = 4, Vf = 1.8 Hz. (with square rooter).

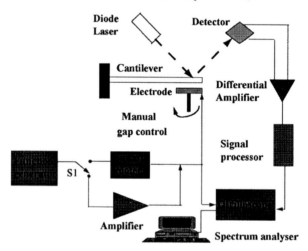

Fig. 4   Experimental arrangement.

in-built dc offset capability was used to provide the drive field and a computational integrated circuit was used to generate the square root of this drive field. The dynamic behaviour of the cantilever with and without the square rooter was measured using optical beam deflection.[5]

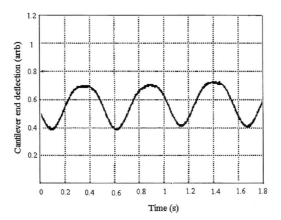

Fig. 5   Cantilever dynamic deflection (dc = 4 V, ac = 2 V).

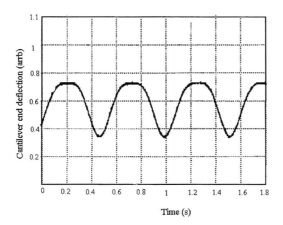

Fig. 6   Cantilever dynamic deflection (dc = 4 V, ac

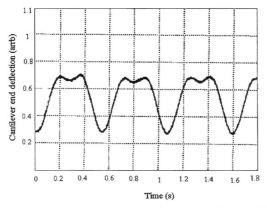

Fig. 7   Cantilever dynamic deflection (dc = 4 V, ac = 6 V).

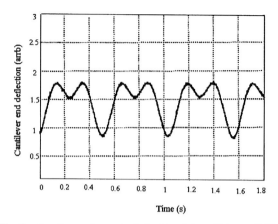

Fig. 8   Cantilever dynamic deflection (dc = 4 V, ac = 8 V).

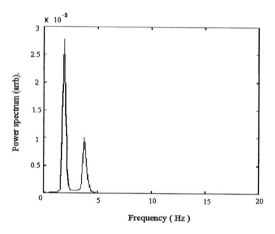

Fig. 9   Power Spectrum of deflection cantilever signal (dc = 4 V and ac = 8 V).

3.1 EXPERIMENTAL RESULTS.

The results obtained without the square rooter are shown in Figs 5–8. In this work the dc bias is held constant while the ac drive field is varied. As can be seen the cantilever response become increasingly non-linear as the ac drive level increased. Figure 9 shows that there is a higher order frequency component generated at twice the drive frequency.

Figures 10 and 11 show that a linear response can be produced by using a square root circuit as expected from the theoretical model. The effect of the square root operation can be clearly seen from the spectrum in Fig. 11. This figure shows that only the excitation frequency is produced.

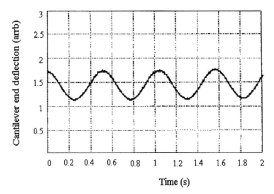

Fig. 10   The cantilever deflection with the square rooter (Sinewave drive signal).

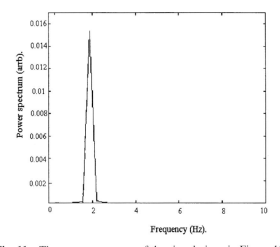

Fig. 11   The power spectrum of the signal given in Figure 10.

CONCLUSION

The performance of the electrostatic actuator with applied dc and ac voltage is theoreti-cally and experimentally demonstrated. This study is carried out to investigate the abil-ity of this type of actuation and its use in micropositioning applications. The results produced have shown the agreement between the theoretical and experimental data is very good. The non-linear characteristic was also investigated and it was demonstrated that an electrostatic actuator can be linearised by means of a square root circuit.

REFERENCES

1.   D. Wood, J.S. Burdess and A.J. Harris: 'Actuators and their mechanisms in microengineering', *IEE digest no: 96/110*, 1996.

2.   J.W. Gardner and H.T. Hingle: 'Developments in Nanotechnology', Volume 1, *Instrumentation to Nanotechnology*, 301.
3.   H.C. Nathhanson, W.E. Newell, R.A. Wickstrom and J.R. Davis: 'The Resonant Gate Transistor', *IEEE Transactions on Electron Devices*, March1967, **14**(3).
4.   M.M. Bakush: *PhD Thesis*, University of Manchester, 1996.
5.   M.J. Cunningham, D.F.L. Jenkins, W.W. Clegg and M.M. Bakush: 'Active Vibration Control of a Small Cantilever Actuator', *Sensors and Actuators A (Physical) Micromechanics*, 1995, A50, *147–150*.

MAGNETOSTRICTIVE MATERIALS AND DEVICES

M.R.J. GIBBS*, R. WATTS, W. KARL, H.J. HATTON,
A.L. POWELL, R.B. YATES AND C.R. WHITEHOUSE

*Department of Physics, University of Sheffield, Sheffield S3 7RH
Sheffield Centre for Advanced Magnetic Materials and Devices, Sheffield
Microelectromechanical Systems Unit, University of Sheffield, Sheffield S3 7RH

ABSTRACT

We discuss the significance of recent advances in magnetostrictive materials, in particular emphasising the advantages to be gained from their integration in microelectromechanical systems. By way of illustration a demonstrator pressure sensor is described.

INTRODUCTION

Magnetostriction is a phenomenon common to all ferromagnetic materials. In the presence of an externally applied magnetic field, H, magnetostriction results in a dimensional change of the ferromagnet (the Joule effect). The dimensional change, λ_e, may be an increase or decrease in length, and may vary in magnitude from parts in 10^7 to parts in 10^3, depending on sample composition and history. It is the Joule effect which may be considered for actuation. A typical magnetostrictive response for an optimised soft amorphous ferromagnetic material is illustrated in Fig. 1. It is important to note the sharp curvature close to zero applied field, the almost linear response over an intermediate field region, and the absence of significant hysteresis. This response is close to the ideal theoretical behaviour.

There is also the inverse effect, where the application of mechanical strain, ε, acts to change the permeability, μ, of the ferromagnet (the Villari effect). It is the Villari effect which may be used for sensing.

There has been considerable development in magnetostrictive materials in recent years. Amorphous ferromagnetic materials, where the production process eliminates long range structural order, have been shown to exhibit the almost ideal behaviour as exemplified in Fig. 1.[1] Problems have, however, been encountered in placing such ribbon-form materials in devices, as bonding stresses can degrade the performance.[2] Nevertheless, very high figures of merit have been reported in device configurations where such problems have been minimised by using a dynamic bond with inherent poor dc performance. The bonding problem might be alleviated by direct deposition of the sensing material onto the test structure.

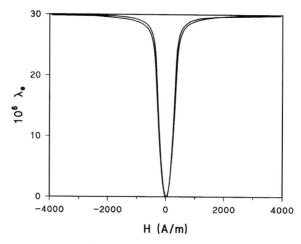

Fig. 1   Magnetostrictive strain, λ_e, against applied magnetic field, H, for an amorphous soft ferromagnet. The sample is field annealed $Fe_{40}Ni_{40}B_{20}$ Further details are contained in Ref. 1.

For a simple strain gauge, the figure of merit, F, for a dynamically bonded ribbon-form Fe–Si–B–C amorphous alloy was measured to be[2]

$$F = \frac{1}{\mu}\frac{d\mu}{d\varepsilon} = 2 \times 10^5 \qquad (1)$$

This should be compared with the corresponding figure of merit for a p-doped silicon piezoresistive strain gauge of around 200.

The highest magnetostrictive strains are found in ferromagnets containing rare earth elements, in particular Tb. However, the differential or peak strain in these materials relies on the generation of large fields for biasing or driving the material. This stems from the large magnetocrystalline anisotropy associated with the rare earth elements. The comparison between materials with and without rare earth elements is reflected in the piezomagnetic coefficients, d_{33}^{max} which are 57 x 10^{-9} mA^{-1} for Td–Dy–Fe compared to 1000 x 10^{-9} mA^{-1} for amorphous Fe–based material.[3]

Most recently, there has been the development of high magnetostriction magnetic multilayers,[4] where the presence of interfaces within the structure appears to give increased magnetostriction constants without a significant loss of magnetic softness. It is in thin film and multilayer materials that we may expect to see the greatest scale of advance in the near future.

In this paper a review is given of new classes of thin film and multilayer material suitable for application to microelectromechanical systems. To discuss the advantages of such a route, a demonstrator pressure sensor is described.

KEY ELEMENTS OF MAGNETOSTRICTIVE BEHAVIOUR

For illustration, we consider only isotropic materials; for non-isotropic materials the physics is identical but the maths is less tractable. The basic material parameter of interest is the saturation magnetostriction constant, λ_s, which for transition metal based ferromagnets is typically 30×10^{-6}. If a magnetic field is applied at an angle θ to the initial direction of magnetisation, causing a uniform rotation of the magnetisation within the sample, then the maximum magnetostrictive strain, λ_e, in the direction of the field (sometimes called the engineering magnetostriction) is given by

$$\lambda_e = \frac{3}{2}\lambda_s\left(\cos^2\theta - \frac{1}{3}\right)$$

(2)

At low fields, it is straightforward to show that under the same conditions, the magnetostrictive strain has a quadratic dependence on the magnetisation, M (which in the ideal case of constant susceptibility is equivalent to H) (see Fig. 1).[1]

The coupling of mechanical strain with the direction of magnetisation also produces the ΔE effect, whereby the apparent Young's modulus is field dependent. Data illustrating this phenomena are given in Fig. 2.

It is these two phenomena which form the basis of sensing and actuation in microelectromechanical systems based on magnetostrictive materials.

AMORPHOUS FERROMAGNETIC THIN FILMS

Rapidly solidified amorphous ferromagnets based on transition elements and rare earth elements have been widely studied, but more recently there have been advances in their production in thin film form.[5,6] Using sputtering it has been possible to produce films with properties very close to the parent rapidly solidified material. Careful choice of sputtering pressure and power has led to successful growth on a range of substrates including Si, Si_xN_y and GaAs. It has also proved possible to use lithographic techniques to pattern these materials.[7] In all cases the growth parameters are adjusted to ensure, as far as is mutually compatible, stoichiometric growth from the target and low stress in the film. Coercivities as low as 15 Am^{-1} have been achieved with $\lambda_s = 35 \times 10^6$ for an Fe–Si–B–C material.

MAGNETIC MULTILAYERS

Based on the electronic changes brought about at surfaces or interfaces, it has been proposed that multilayers formed of transition metals and noble metals may demonstrate enhanced magnetostriction constants compared to bulk materials.[4,8] We have particularly studied $Fe_{50}Co_{50}/Ag$ multilayers. This FeCo alloy was chosen as it has the highest

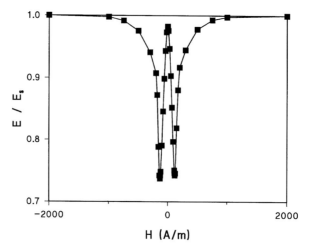

Fig. 2   The ratio of Young's modulus, E, to that at magnetic saturation, E_s, as a function of applied magnetic field, H, for an amorphous soft ferromagnet. The sample is field annealed $Fe_{40}Ni_{40}B_{20}$ Further details are contained in Ref. 1.

saturation magnetostriction constants of any non-rare-earth based material, and the Ag as it is immiscible in FeCo and has a lattice mismatch to it. Figure 3 gives some illustrative data for this system, which may be interpreted as demonstrating, from the non-zero intercept in the upper plot, a significant contribution to the total magnetostriction from the Ag/FeCo interfaces. Again, careful choice of sputtering power and pressure has allowed us to move towards optimum conditions for this system. The lower plot in Fig. 3 demonstrates the dependence of the interface contribution on sputter deposition parameters.

INTEGRATION IN MICROELECTROMECHANICAL SYSTEMS

A microelectromechanical system is, quite generally, an intelligent system comprising sensing, processing and/or actuating functions, integrated onto a single chip or multichip hybrid. We have already published data on an optically interrogated, magnetically tuned, micro-resonator.[9] Other obvious areas of application include the recording head in a magnetic data storage system, where there are micromachined components, which may include microactuators.[10]

A number of materials issues have to be addressed for successful integration. The substrate material must be chosen in order to meet the overall demands of the device in terms of mechanical and environmental integrity. With the expertise developed in the semiconductor industry, this usually comes down to a choice between Si and GaAs. For the design and fabrication of micromechanical sensors and actuators, material properties need to be known very accurately. The working principles of these devices depend strongly

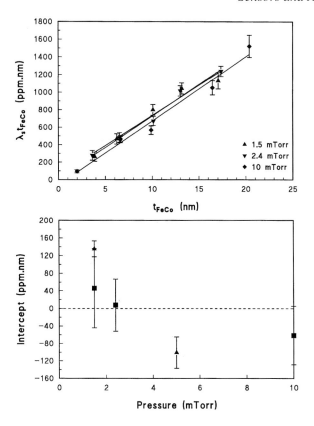

Fig. 3   The top figure shows saturation magnetostriction data for as-deposited multilayers of the form 2 x 25(2nmAg/XnmEe$_{50}$Co$_{50}$) on polyimide substrates. The data is interpreted in terms of the model in Ref. 4. Further details may be found in Ref. 7. The bottom figure is the derived contribution from the interfaces in the multilayer, according to the model in Ref. 8.

on the quality and control of the used materials.

Whilst mechanical, thermal and electrical data may be well known for bulk forms of these materials, data on thin film and micromachined components still appear scattered. Most essential is the knowledge of the elastic modulus E of the thin film material. The Young's modulus of chemical vapour deposited (CVD) silicon nitride films, often used in pressure sensing elements, is reported to be within a range of 100 GPa to 330 GPa.[11–15] The large differences in the literature can be attributed to different processing conditions and postprocessing treatments. It is therefore necessary to determine the exact values for specific applications. For microstructures coated with magnetic materials and exploiting magnetoelastic effects, the Young's modulus of the composite structure is of interest. In house measurement methods for the characterisation of mechanical thin film properties are currently being developed (see later).

There may be a significant thermal mismatch between the metallisation and the sub-structure. If any stage of the fabrication or deployment of the device involves elevated temperature, this will be a significant factor in the overall performance. The integrity of the interface between the metallisation and substructure must be investigated. If the device is to see many mechanical duty cycles before failure, the interface must be free from such problems as delamination. Account must be taken of any cross-contamination between the magnetic film and the substructure, such as diffusion of elements producing a significant change in magnetic properties.[16] The steps involved in defining the device structure may include exposure to temperature, etchants and ion bombardment. All of these steps must be examined for their overall effect on final performance.

INCREASED FUNCTIONALITY FROM INTEGRATION OF MAGNETIC FILMS IN MICROELECTROMECHANICAL SYSTEMS

Notwithstanding the issues discussed above, there are a number of significant benefits to be derived from pursuing such devices. We illustrate these by example.

If a simple cantilever is taken as a basic accelerometer, methods have to be found to calibrate, test and interrogate the device. The choice is between electrical and magnetic methods. Electrical transport methods (e.g. piezoresistive effects), and capacitance methods rely on direct connections to the cantilever. This usually involves extra metallisation stages to define the contacts. Controlled mechanical deflection can then be used for calibration. Self test is possible using the capacitance method and an electrostatic deflection, but cannot be achieved for the piezoresistive case. Interrogation comes by monitoring the signal (resistance or capacitance) over time. The device is active in the sense that current must be flowing at all times in the piezoresistive case.

If magnetostrictive effects are used the situation is different. We have already pointed out the substantially enhanced figure of merit for the magnetostrictive (piezomagnetic) case. If the simple cantilever is coated with piezomagnetic material then no direct contacts need to be made as Faraday's law allows coupling between a coil and the magnetic material. We have modelled this situation for coated membranes,[17] and clearly see the benefit of the piezomagnetic response. Calibration comes from the application of a known field, and the detection of deflection at the end of the cantilever by means of the Villari effect. Self test is possible by the same method. If the remote coil is excited at a suitable frequency, the coil-film combination acts as a cored inductor. The impedance will change as the Villari effect operates on deflection of the cantilever. In all cases there is no direct contact with the active sensing element. This could allow for environmental packaging. One metallisation at the sensor head covers all required functions.

DEMONSTRATOR PRESSURE SENSOR

In order to address a number of the issues raised above, we are developing a proof of principle pressure sensor. This is based on a closed magnetic circuit, where one part of the circuit overlays a micromachined membrane. A coil is microfabricated to couple to the magnetic circuit in such a way that the impedance of the coil can be measured as the membrane is deflected by pressure changes.

The magnetic material chosen is amorphous Fe–Si–B–C alloy derived from a METGLAS® 26055C target. This is grown onto GaAs using rf magnetron sputtering to a thickness of 1 μm. The dc coercivity was 60 A/m for the unpatterned film.

The prototype device was fabricated using an undoped semi-insulating (UDSI) GaAs wafer as the substrate, (predominately because of the in-house expertise in the processing of this material system). The lower coil metallisation was recessed into the surface of the substrate by evaporating the Au conductors into a pre-etched channel. The surface was then coated with a 0.3 μm layer of PECVD Si_3N_4 to provide both insulation between the coil and magnetic material, and the base structure from which the free standing membranes will be constructed.

A 1μm thick magnetic film was rf sputtered over the entire surface of the sample and patterned using standard photolithography and a wet chemical etch to give a sloping sidewall profile. The sample was once again coated with a 0.3μm thick layer of Si_3N_4 to insulate the top surface of the magnetic material. Via holes were plasma etched through both Si_3N_4 layers to enable the embedded lower coil metallisation to be contacted. The upper coil metallisation was evaporated on to the sample and patterned photolithographically using a lift-off technique. Preliminary results demonstrate the necessary transformer action between sets of coils. We are now proceeding to open out the membrane at the rear of one part of the magnetic circuit.

To model sensor operation, the stress distribution over the membrane area needs to be determined. For a maximum change in permeability a region of uniform, uniaxial, high stress or strain is preferable. FEM simulations using ANSYS® showed that for this purpose high aspect ratio, rectangular membranes seem to be most suitable. Based on the known stress distribution in the material the change of its magnetic properties and therefore the change in inductance/impedance of the magnetic circuit can be derived.

To model the electromagnetic behaviour of the membrane, it is important that the mechanical properties are well understood. As noted previously, the Young's modulus of a thin film cannot be simply assumed to be that of the bulk material. We have developed a method for measuring the deflection of a membrane as a function of the pressure difference across it. The deflection is measured using a sodium lamp interferometer, with each fringe spacing corresponding to one wavelength change in the total light path length (half a wavelength deflection of the surface). An example measurement is shown in Fig. 4. The pressure difference is provided by a head of water, which also gives a simple and accurate direct measurement of the pressure (1 mm of water corresponds to approximately 10 Pa).

Using the formula for central deflection of a stressed circular membrane the data can be linearised. From the gradient and intercept, the Young's modulus and intrinsic stress in the film can be independently calculated. For a 250 nm thick SiN membrane on GaAs we obtained E = 157 GPa with an intrinsic stress of 12.3 GPa.

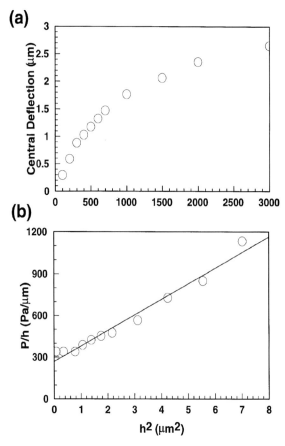

Fig. 4   (a) Central deflection of a circular Si$_x$N$_y$ membrane of radius 200 μm and thickness 250 nm as a function of pressure, (b) linearised plot of the same data – the Young's modulus can be determined from the gradient and the intrinsic stress from the intercept.[13]

CONCLUSIONS

We have demonstrated that piezomagnetic coatings can provide useful increased functionality in microelectromechanical systems. Fabrication of a proof of principle pressure sensor has been described, together with preliminary characterisation.

ACKNOWLEDGEMENTS

This work is supported by the UK Engineering and Physical Sciences Research Council under its Advanced Magnetics Programme. We also acknowledge Alan Walker and Paul Haines in the Departmental Clean room for their fabrication work.

REFERENCES

1. A.P. Thomas and M.R.J. Gibbs: 'Anisotropy and magnetostriction in metallic glasses', *J. Magn. Magn. Mat.*, 1992, **103,** 97–110.
2. M. Wun-Fogle, H.T. Savage and A.E. Clark: 'Sensitive, wide frequency range magnetostrictive strain gage', *Sensors and Actuators*, 1987, **12,** 323–331.
3. E. du Trémolet de Lacheisserie: *Magnetostriction, theory and applications of magnetoelasticity*, CRC Press, London, 1993.
4. T.A. Lafford and M.R.J. Gibbs: 'Interpretation of saturation magnetostriction data in multilayer systems', *IEEE Trans. Mag.*, 1995, **31**(6), 4094–4096.
5. A.D. Mattingley, C. Shearwood and M.R.J. Gibbs: 'Magnetic and magnetoelastic properties of amorphous Fe-Si-B-C films', *IEEE Trans. Mag.*, 1994, **30**(6), 4806–4808.
6. E. Quandt and K. Seemann: 'Fabrication of giant magnetostrictive thin film actuators', *Proc. IEEE Micro Electro Mechanical Systems MEMS'95*, Amsterdam, 1995, 273–277.
7. C. Shearwood, M.R.J. Gibbs and A.D. Mattingley: 'Growth and patterning of amorphous FeSiBC films', *J. Magn. Magn. Mat.*, 1996, **162,** 147–154.
8. H.J. Hatton and M.R.J. Gibbs: 'Interface contributions to magnetostriction in transition metal based soft magnetic multilayers', *J. Magn. Magn. Mat.*, 1996, **156,** 67–68.
9. M.R.J. Gibbs, C. Shearwood, J.L. Dancaster, P.E.M. Frere and A.J. Jacobs-Cook: 'Piezomagnetic tuning of a micromachined resonator', *IEEE Trans.Mag.*, 1996, **32** 4950–4952.
10. W. Tang, V. Temesvary, R. Miller, A. Desai, Y.-C. Tai and D.K. Miu: 'Silicon micromachined electromagnetic microactuator for rigid disk drives', *IEEE Trans.Mag.*, 1995, **31**(6), 2964–2966.
11. J.J. Vlassak and W.D. Nix: 'A new bulge test technique for the determination of Young's modulus and Poisson's ratio of thin films', *J. Mater. Res.*, 1992, **7**(12), 3242–3249.
12. D. Maier-Schneider, J. Maibach, E. Obermeier: 'Computer-aided characterisation of the elastic properties of thin films', *J. Micromech. Microeng.* 1992, **2,** 173–175.
13. E.I. Bromley, J.N. Randall, D.C. Flanders and R.W.Mountain: 'A technique for the determination of stress in thin films', *J. Vac. Sci. Technol.* 1983, **B1**, 1364–1366.
14. K.E. Peterson: 'Silicon as a Mechanical Material', *Proc. of the IEEE*, 1982, **70**, 420–457.
15. J.A. Taylor: 'The mechanical properties and microstructure of plasma enhanced chemical vapor deposited silicon nitride thin films', *J.Vac.Sci.Technol.*, 1991, **A9,** 2464–2468.
16. J.A.C. Bland, R.D. Bateson, P.C. Riedi, R.G. Graham, H.J. Lauter, J. Penfold and C. Shackleton: 'Magnetic properties of bcc Co films', *J. Appl. Phys.*, 1991, **69**, 4989–4991.
17. W. Affane, M.R.J. Gibbs and A.L. Powell: 'Performance modelling of micromachined sensor membranes coated with piezomagnetic material', *Sensors and Actuators*, 1996, **A51**, 219–224.

MATERIALS SCIENCE IN ER FLUIDS AND DEVICES

W.A. BULLOUGH

*'Smart Machines, Materials and Related Topics', University of Sheffield,
Department of Mechanical Engineering, Mappin Street, Sheffield, S1 3JD.*

ABSTRACT

Electro-rheological fluids are reviewed with respect to their application in devices that are aimed at featuring electronically designated motion and flexible operation - 3rd wave machines. The article is one engineer's view of what is required in order to promote a successful and unified approach to the interdisciplinary problem amongst scientists and engineers. This is done by setting down the state of the art position of research in the field and the salient factors that determine research trends for fluid developers, whilst at the same time giving some idea of desired machine performance and limitations.

INTRODUCTION

Current interest in electro-rheological fluids (ERF) arises principally as a bi-product of the search for high speed, flexibly operated light payload machines. In these devices the motion is to be electronically programmed and take place without any mechanical re-configuration of the moving parts.[1]

Generally speaking the heavier loads are catered for in so much as they can be subjected to adequate control by servo and stepper motors; there the inertia of the drive is relatively small and the respective response time is short compared to the duration of the acceleration of an external load and all other variations in motion. There probably exists an area of light loads in which the necessity of rapid 'take off' after signal and intense acceleration thereafter cannot be achieved by magnetic means.

The term machine, as used in this article, implies a certain range(s) of useful force, displacement and velocity. A major limitation to machine performance is the generation of heat and its transfer. Since lubrication and mass transfer are usually associated with liquids and in the light of the electronic requirement the search is on for a liquid state equivalent to the solid state semiconductor or (see later) transistorised fluid.

The pressing home of this background philosophy and its linking to material science requirements is the central theme of this article. Without this the possibility that engineering and scientific endeavours may fail to coincide is too strong.

SCOPE OF TARGETS FOR ERF

Taking realistic values for light machine duty as say: nominal change of speed = 10 ms^{-1}, sweep = 100 mm with a stopping precision of 0.1 mm within 10 milliseconds of receiving a hold signal then inertial loadings of 100 g acceleration and 0.1% strain are arrived at. This is a region where the selection of materials for belts, pulleys etc reaches a threshold. Certainly at 1000 g the critical length for materials (on both yield stress and maximum strain limitations/considerations) is too limiting for all but a few materials and part sizes.[2]

This scenario seems to give the first and most important pointer to the required performance of ERF, i.e. it is pointless to search for properties in the fluid which could not be utilised on the machine. At 100 g acceleration and at realistic machine part sizes the disposition (shape) of material is already a problem, as is fluid temperature control for the nominal motion (see also later section on fluid inertia/super slip). Thus 100 g is taken as the target. With the duties stated the task can now be made clear in a quantitative sense.

A further and most important feature of the ERF/flexible high speed machine scenario is that the optimisation problem is complex. To achieve the target in a reliability sense requires integration in disciplines concerning materials, electronics, chemistry, heat transfer etc. It is the authors contention that only by initially understanding the forgoing situations can material and application requirements be appreciated.

The time for materials scientists per se to become further involved in ERF is ripe. Targets in a dynamics sense are on the verge of being achieved. Reliability and serviceability will surely soon replace the call for more potency (10^{+} kPa yield is adequate). What follows, is a restatement of the problem and related parameters on which to quantify it, it is for material scientists and not a mechanical engineer to provide the materials answers and to review, describe and perfect the modus operandi.

ER MODELS AND CHARACTERISATION

Perhaps the most problematic area which has obstructed the furtherance of electrorheological fluids is the particle mechanics or continuum conundrum for characterising the flow of the dense semi conducting slurry (unexcited) or yielding plastic (excited). Whilst many years have been spent in computer analyses based on dielectric mismatch polarisation alone, models of the multi-dielectric bodied excited structure the yield stress levels were under predicted from the constituents by an order of magnitude. On the other hand it is not difficult to find texts that claim that the performance of the same fluid in Couette-shear flow and Poiseuille valve flow cannot be related i.e. the fluid cannot be treated as a continuum.[3]

It is now evident that polarisation is not the only mechanism (Fig. 1) at work and that disparate liquid/solid conductivity[4–7] plus hydrodynamic effects – at least[8] need to be included in rheological models designed to illustrate the modus operandi of the effect

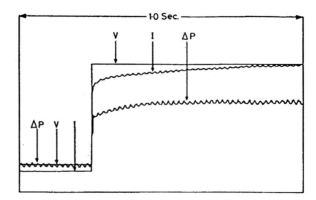

Fig. 1   Conductivity is seen to be important in the secondary response (at least). Steady flow in valve with step voltage V applied. ΔP is valve pressure drop and I the electrical current. Similar behaviour is seen in a clutch.

and to link it quantitatively to fluid properties, flow conditions and excitation levels, i.e. with some accuracy. The possibilities for continuum means of characterising ER fluids in flow (otherwise the properties of viscosity and density have little significance and design techniques become entirely empirical) have been given a boost by work which indicates a simple approximate link between the fluids performance in the above modes of flow and others (static shear) for different fluids. This is done by the use of non-dimensional Hedström and Reynolds Numbers via use of the Buckinghams relationships for a Bingham Plastic[9] – for steady flow (Fig. 2). This is welcome yet perhaps surprising since the thickness of the shearing fluid layer near a boundary can approximate to a particle (often of variable geometry) size.

The extended concentration by researchers on the slow steady flow detail belies the necessity to confront the intensely unsteady (and indeed steady high shear rate motions) that will be required in practical machine work cycles. Very often misleading appraisals of situations can arise from the assumption of quasi-steady flow, due in part to the neglect of severe inertial or compressibility effects. Also, correspondence between shear and flow mode time domain performance has not been shown across the shear rate range.

3RD WAVE MACHINES

In a high speed flexible machine – there is little scope for the generation say of motion as by an inductive/heavy rotor electro-magnetic drive or the by generation of a shaped control voltage. In both cases respectively, latching onto a high inertia source of steady motion (and the braking of it) and a high capacity, high tension supply line (and discharge by earth shorting) produces a digital event.[10] Both AC and DC excitation components are present in step switching and dwell periods respectively yet, there is a tendency to separate fluids into AC or DC types. The supposition is growing that potent AC

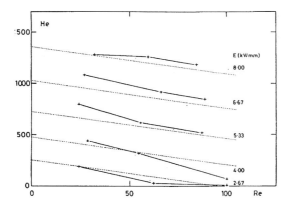

Fig. 2   Hedstrom Number versus Reynolds Number for a valve (full line experimental result) as predicted from clutch (Couette mode) expenemental data (dotted) for different values of V/h = E..

fluids depend on particle polarisation in a non conductive fluid whilst apparently good DC fluids need current flow – both based on only steady state strength and current flow appraisal.

Returning to the ultra high acceleration/low inertia flexible machine regime, it is predicted that the limiting change of speed response time t_o, in the digital mode of operation is determined by a factor proportional to $h^2\rho/\mu$ which is not a function of τ_e the fluids yield stress but is heavily influenced by the inter electrode gap size h, and the fluids density ρ and viscosity μ[11, 12] (Fig. 3). This is important when the fluid self load, rather than the inertia of the solid part power transmission load dominates the mechanics. It is a somewhat contrary situation to the requirement for the control of the fluid viscous heating problem.

For a change speed response time less than say 10 m sec a 4×10^5 V/S signal rise/fall time is required. This has implications for the fluid capacitance which is not simply modelled as a function of shear rate.[13] When the voltage is rapidly applied the yield shear stress follows at a time constant of approximately RC, the resistance and capacitance product of the inter electrode space. There is little point in accelerating the load rapidly if the torque initiation lags much behind the steep change of excitation by t* – this (alas) can be difficult to measure.[14, 15] Fortunately the lag seems to decrease the harder the fluid is being punished in terms of E, $\dot{\gamma}$, 0 (electric field, shear rate and temperature) respectively (Fig. 4).

This factor becomes important if the generation of a motion profile in a 3rd wave machine is considered. Without getting involved with digital technology: if the x direction speed provided is constant then the y penetration (driven by a bang–bang application of voltage needs a yield stress of sufficient magnitude to give the relevant part high and instant acceleration) must be maintained over a very small time interval (fixed by the switching speed) if the resolution is not to be too crude. DC operation seems virtu-

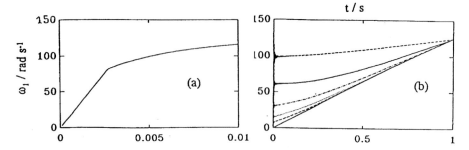

Fig. 3 Analytical prediction based on Bingham Plastic constitutive equation of lightly loaded cylindrical clutch performance (a) limiting run up time of output rotor speed (b) super slip of velocity profile between electrodes. r is a general radius with R_1 the radius of the inner driven member. co is the speed of the driven member at time t after switch on of step voltage between constant speed driving memebr and driven member electrodes. K depends on geometry of device.

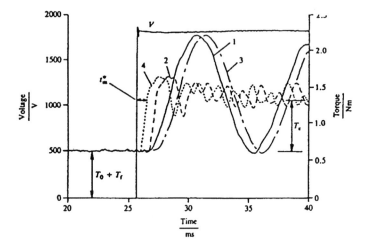

Fig. 4 Numerical transformation of step torque h 0.5 mm; 1. Measured torque-transducer signal; 2. First estimate of ER clutch torque response; 3. Predicted torque-transducer response for first estimate; 4. Final estimate of ER clutch torque response + Tf are viscous and real friction torques with **Te** due to application of step voltage V.

ally mandatory with any hysteretic and electrophoretic tendencies being arrested by a conjunction of binary switching and high $\dot\gamma$ (Fig. 5).

Because of this concept plus hysteresis effects description of the steady state form of current i and τ_e on field strength are not dwelt upon. Both vary between proportionally to E^2 and $E-E_0$, (where E_0 is the lower voltage level deadband) depending on the particular fluid. Temperature in an important non linear factor in both relationships and $\dot\gamma$ less so; when voltage and temperature are near optimum $\dot\gamma$ need not be a large factor in τ_e.

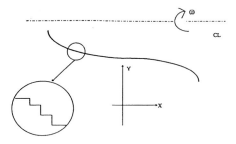

Fig. 5   Digital ER motion synthesizer (concept); y direction is shown ER controlled with x direction of equal time elements.

MECHATRONICS AND TESTING

It does not seem possible to provide a figure of merit for a fluid which possesses these sundry needs but, the linear traverse mechanism will demonstrably test total capability in that respect.[16] In this device two contra-rotating high inertia, constant velocity rotors provide motion sources with HT and earth 'busbars' the excitation via switches. Two driven clutches spaced from their driver co axially by the ERF, are connected to a pulley each of which is connected by a common belt. The ERF in opposing clutch drives (Fig. 6) is excited alternatively to make the belt reciprocate, with typical steady speed of up to ± 5 ms^{-1} separated by turn round times determined by the fluid properties: τ_e, μ, θ and t* are involved in the turn round performance; μ can also distort the traverse profile or prevent take up, if excessive. A top quality fluid should turn round in circa 20 m sec; thermal run away (the liquid state has a negative temperature coefficient) should be avoided by a large margin as possible; the heat transfer rate from the outer driving rotor is about the maximum per unit area that is achievable into the atmosphere. The electron-hydraulic time delay t* will be typically 1 msec. The full speed centripetal field on the particle is about 300 g and the belt acceleration 100 to 200 g. Strangely this does not necessarily cause rapid fluid degradation – if careful design procedures are followed.[17, 18]

In analysing such performance data the fundamental compatibility between a low time constant, the heating effect of the viscous shearing and resistive loads and the level of voltage arises. Alas, the failure of fluid on this machine implies its separate analysis on each of several simple characteristics tests – in order to isolate the problem area. The test does however give a good example of the machine side of the overall electrical-chemical-rheological thermal fluid/machine optimisation. It will be appreciated that the inertia, geometry and stress and strain in mechanical parts are linked to operating conditions, and particularly to acceleration in the unsteady mode. For example, the uniform speed of the traverse could be obtained (presumably) by having a long, small radius clutch and having a large pulley and a low rotational speed. Likewise, fluid performance τ_e, t*, μ depends on the solids content, materials properties, size and shape of particles, θ and $\dot{\gamma}$. With present lowish fluid yield strength properties the optimisation process is

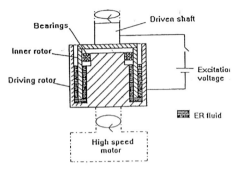

Fig. 6   Cylinderical clutch for ER traverse gear. Pullys on driven shafts (contra-rotating) pulleys are connected by belt. Alternative excitation of clutches causes reciprocation of belt which carries product to be wound on a bobbin (not shown).

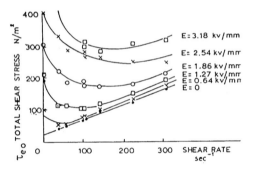

Fig. 7   Typical rising τ v $\dot{\gamma}$ steady state clutch characteristic showing ofien encountered low shear rate instability tendency zone.

made more hit and miss if extensive fluid data is not available.[19]

In connection with the traverse mechanism e.g. the need for a yield stress and a rising τ characteristic with $\dot{\gamma}$ is noted (Fig. 7). This is necessary for any clutch drive where an overload may for example cause slip. If τ_e was then to fall, and hence the torque, further slip and stall would occur. However, the Bingham Plastic characteristic per se is not obligatory. In other types of flexible machines such as the vibration isolator, a purely viscous force/velocity characteristic seems preferable;[20] homogeneous liquid crystal type fluids could become important here.

In the flow mode of operation much the same factors come into play as in the shear mode, save that the ER fluid can be passed out of the system for cooling and the residence time of fluid between valve plates is short. Inertial/acceleration pressures can dominate the total pressure magnitude, especially in a vigorous valve controlled piston/cylinders arrangement[14] – as in a shock absorber or vibrator. The ratio of piston area to its control valve size is an important variable in both this context and the magnification of yield stress into piston force. Compressibility effects are, as per normal hydraulic practice, a limitation principally on pressure rise time after excitation; fluid inertia and compressibility effects should not be too important below an operating frequency of 100 Hz.[21] The benefits of a high yield stress magnitude is to reduce the amount of fluid volume for a given force requirement and hence improve the response time as well as being generally beneficial in a mechanical sense. When the above factors do not dominate the problem a phenomenon similar to the super slip that occurs in a lightly loaded accelerating clutch limits operation.[12]

HYSTERESIS AND CONTROL

The valve controlled application in general exemplifies an interesting control problem. Whilst good reasons have been given for digital control of a particulate ERF the damper could be envisaged as a continuous ride member under analogue excitation/control. Past studies have however shown a pseudo or perhaps time dependent hysteresis[22, 23] which is better treated by on-off operation, otherwise more than a suggestion is apparent that voltage alone is insufficient as a control parameter.[12] These and sometimes experienced violent clutch (shuddering) and valve (choking)[24] operation may yet prove not to be separate phenomena and all are possibly linked and related to structure formation. These effects plus electro phoresis are to be avoided save for their further investigation (Fig. 8).

Shear modulus G^1, and C_v specific heat capacity and bulk modulus K under field need investigation since they can also determine the precision of any controlled positioning device.

LIMIT OF CONCEPT

All of the foreseen effects put a limit an the performance of a flexibly operated machine and set the requirements for τ_e and $t^* f (\dot{\gamma}, \theta, E)$ in the ER fluid. There may be a competing limiting factors – fluid elasticity and volume, super slip, heating and cooling etc. Further limitations arise from lubrication – e.g. particles will only move through an elasto-hydrodynamic region at low speeds (Fig. 9) and anti-wear boundary lubricity is hence very important.[25] In film bearings of the hydrostatic Rayleigh type the time constant appears to be inordinately slow.[26] Other limitations arise from the lack of τ_e, t^* materials algorithms with respect to particle concentration and size, dielectric and conductance etc properties.

The self weight/inertial loading problem[2] can easily be avoided so far as solid material critical breaking length in concerned but strain will limit the overall acceleration (on grounds of precision) – only a few material will exhibit say less than 0.01% strain at an acceleration of 1000 g. Accelerations of l00 g+ are regularly attained in machines and cause one to wonder at the rate of separation of particles and possible cavitation effects in the fluid.

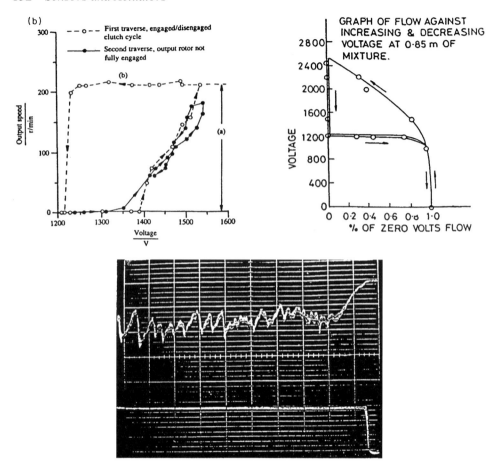

Fig. 8 Hysteresis/structure related effects in (a) Couette viscometer (b) clutch in on-off DC operation (c) valve experiencing choking phenomena at nominally constant flow rate -upper trace is ongoing DC voltage and lower, valve pressure drop versus time (d) valve fed by header tank-flow stops and starts at different voltages; hysteresis is not noticeable without a stoppage.

FUTURE

The subject of high speed, flexibly operated electronically reconfigureable machines based on ER fluids is intensely multi disciplinary and highly non linear in terms of analysis, and the limits of operation cannot be graphically represented – such is the degree of compromise required between fluid design, motion and machine. An article of this type cannot give comprehensive treatment of the interfacial problems, rather it examplifies and lists the more apparent and important factors. Having done this it is hoped that the targets to suit fluid developers can be set more effectively than hitherto – see appendix.

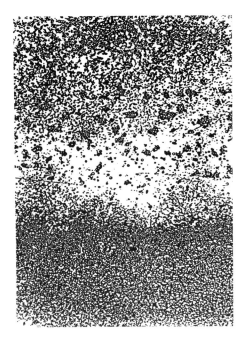

Fig. 9   Tracks left on the surface of a flat specimen after the passage of a belt rolling at (a) high speed 50 mm/s and (b) low speed. At high speed particles are unable to enter the contact between the rolling ball and plate surface; a depleted track is observed. At low speeds the particles enter the contact and have a lubricating effect. (Steel ball diameter 25mm rolling on a glass disk. Optical interferometic apparatus).

Finally a word of warning to all potential researchers: a τ_e v $\dot{\gamma}$ characteristic for the fluid will not give exactly the same shape of torque v speed or pressure v flowrate curves for a device, and viscometers shoud be desigued and operated so that the fluid rather than device characteristic is measured[27] – in steady and time domain in other areas requiring specialist attention include lubrication, hysteresis, stabilisation and water exclusion.

The slower, more heavy duty applications of magneto rheological structured fluids are now beginning to appear commercially. The higher yield stresses available from this medium have recently brought announcements of commercially available excavator cab dampers, a programmable exercise machine and flexibly operated litho lens grinders in which the controlled fluid itself abraids.[1] These devices are setting the real target for the remaining phase of 3rd Wave of Machines viz the fast domain of lightweight controlled devices at which ERF is aimed. A comprehensive survey of ER fluids technology is given in Ref. 28 and updated in Ref. 29.

The 100 g light duty targets are now being approached in tests but sometimes without the neccesaly fluid life required for the real environment. Whatever testing is done needs to be relevant to the real scale e.g. $0 < \gamma < 20\,000\ \delta^{-1}$, $0 < E < 8$ KVmm^{-1}.

REFERENCES

1. W. A. Bullough: *The Third Wave of Machines*, Endeavour, Pergamon Press, June 1996, 1050–55.
2. J.R. Yates, D.S. Lau and W.A. Bullough: 'Inertial Materials: Perspective, Review and Future Requirements', Proc. AXON-VDI/E Actuator '94 Conference, Bremen, 275–278.
3. H. Janocha, B. Rech and R. Bolter: 'Practice Relevant Aspects of Constructing ER Fluid Actuators', *ibid*, 435–447.
4. M. Whittle, D.J. Peel, R. Firoozian and W.A. Bullough: 'Dependence of ER Response Time on Conductivity and Polarisation Time', *Phys. Rev. E. Pt6A*, 1994, **49**, 5249–5259.
5. C.W. Wu, Y. Chen, X. Tang and H. Conrad: 'Conductivity and Force between particles in a model ERF I-Conductivity II - Force, *Proc 5th mt. Conf. ER Fluids/MR Suspensions*, Sheffield, World Scientific Proc., 1995, 525–536.
6. C. Boissy, P. Atten and J. N. Foulc: 'The Conduction Model of Electro Rheological Effect Revisited', *ibid*, 156–165.
7. L.C. Davies and J.M. Ginder: 'Electrostatic Forces in Electrorheological fluids', *Progress in Electrorheology*, Plenum Press, NYC, 1995, 107–114.
8. J. Melrose, S.I. Itoh and R.C. Ball: 'Simulations of ERF with Hydrodynamic lubrication', *Proc 5th mt. Conf. ER Fluids/MR Suspensions*, Sheffield, World Scientific Press, 1995, 404–410.
9. D.J. Peel, R. Stanway and W.A. Bullough: 'The Generalised Presentation of Valve and Clutch Data for an ER Fluid and Practical Performance Prediction Methodology', *ibid*, 279–290.
10. R. Tozer, C.T. Orrell and W.A. Bullough: 'On-Off Excitation Switch for ER Devices', *Imt. J. Mod. Phys. B.*, 1994, **8**(20 and 21), 3005–3014.
11. M. Whittle, R. Atkin and W.A. Bullough: 'Fluid Dynamic Limitations on the Performance of an ER Clutch', *J. Non Newtonian Fluid Mech.*, 1995, **57**(1)61–81.
12. M. Whittle, R.J. Atkin and W.A. Bullough: 'Dynamics of an Electro Rheological Valve' *Proc 5th mt. Conf. ER Fluids/MR Suspension*, Sheffield, World Scientific Press, 1995, 100–117.
13. A. H. Sianaki, W. A. Bullough, R. Tozer and M. Whittle: 'Experimental Investigation into Electrical Modelling of Electro-rheological Fluid Shear Mode', *Proc. IEE, Sci. Mest. & Tech.*, **141**(6), 531–537.
14. M. Whittle, W.A. Bullough, D.J. Peel and R. Firoozian: 'Electrorheological Dynamics Derived from Pressure Response Experiments in the Flow Mode', *J. Non Newtonian Fluid Mech.*, 1995, **57**(1), 1–25 .
15. W.A. Bullough, J. Makin, A.R. Johnson, R. Firoozian and A.H. Sianaki: 'ERF Shear Mode Characteristics: Volume Fraction, Shear Rate and Time Effects', at press *Trans. ASME, J. Dyn. Syst. Meast. and Control*, 1995.
16. A.R. Johnson, J. Makin, and W.A. Bullough: 'E. R. Catch/Clutch Simulations', *Int. Jnl. Mod. Phys. B.*, 1994, **8**(20 and 21), 2935–2954.

17. A.R. Johnson, J. Makin, and W.A. Bullough: 'ER Catch/Clutch Simulations', *Int. J. Mod. Phys. B.*, **8**(20 and 21), 2935–2954.

18. C. Wolfe and E. Wendt: 'Application of ERF in Hydraulic Systems', *Proc. AXON-VDL/E Actuator '94 Conference*, Bremen, 284–287.

19. W.A. Bullough, J. Makin, and A.R. Johnson: 'Requirements and Targets for ER Fluids in Electrically Flexible High Speed Power Transmission', Am. Chem. Soc. Fall Meeting Washington DC Plenum Pub., 295–302.

20. A.S. Naem, R. Stanway, J.L. Sproston and W.A. Bullough: 'A Strategy for Adaptive Damping in Vehicle Primary Suspension Systems', 1994 Proc. ASME. Winter Annual Meeting, Chicago, 1994, 395–399.

21. D.J. Peel, R. Stanway and W.A. Bullough: 'Engineering with ERF: A Design Methodology based upon Generalised Fluid Data', *Int. Conf. Intel Materials*, Lyon June 1986, at press.

22. A. Hosseini, H. Sianaki, J. Makin, S. Xiao, A.R. Johnson, R. Firoozian and W.A. Bullough: 'Operational Considerations in the Use of an ElectroRheological Catch Device', *Proc. Soc. Fluid Power Transmission and Control, 1st Fluid Power Trans. & Control Symp.*, Beijing Inst. Tech. Press, Beijing, 1991, 591–595.

23. H. Block and J. Kelly: 'Electro-rheology', *J. Phys*, 1988, **D21**, 1661.

24. D.J. Peel and W.A. Bullough: 'The Effect of Flowrate, Excitation Level and Solids Content on the Time Response of an Electro-rheological Valve', *J. Intel. Matl. Systems and Structures*, 1993, **4**(1), 54–64.

25. R. Dwyer-Joyce, W.A. Bullough and S. Lingard. 'Elastohydrodynamic Performance of Unexcited Electro-Rheological Fluids', *Proc. 5th Int. Conf. ER Fluids/MR Suspension*, Sheffield, World Scientific Press, 1995, 376–384.

26. T.H. Leek, S. Lingard, W.A. Bullough and R.J. Atkin: 'The Time Response of an Electrorheological Fluid in a Hydrodynainic Film', *Proc 5th Int. Conf. ER Fluids/ MR Suspension*, Sheffield, World Scientific Press, 1995, 551–563.

27. R. Atkin, Xiao Shi and W.A. Bullough: 'Solutions of the Constitutive Equations for the Flow of an ElectroRheological Fluid in Radial Configurations', *Proc. Soc. Rheology, J. of Rheology*, 1991, **35**, 1441–1461.

28. Electro-rheological Fluids; 'A research needs assesment' US Department of Energy Office of Energy Research. DE. ACO2-91 ER 30172.

29. Proceeding 5th International Conference *ER Fluids/MR Suspensions*, Sheffield, World Scientific Press, 1995.

APPENDIX

QUESTIONS REMAINING ABOUT ER FLUIDS

1.   How far are the form of test results on one fluid applicable to other fluids?

2.   If fluids such as those from Stangroom and Block had a 60% volume fraction what would the yield stress be? On the other hand what influences e.g. the viscosity for a given concentration of mixture?

3. Some fluids are more sensitive to shear rate and temperature than others in their τ, v, γ, t*, v, γ characteristic; why?

4. In general little is known of the unsteady state behaviour of fluids. Is this not a severe handicap when fluids are to be applied to transient machine operation?

5. Secondary properties of fluids are important in engineering design calculations. Should G^1, Cv, K (compressibility) not get more attention in fluid design than the hitherto?

6. Should conductance not be a prime concern in the design of fluids (and also viscosity) as a function of temperature? Heating in closed ERF systems such as clutches is a limitation on machine application if only in the t* range of fast operation – it should always be small compared to t_0.

7. The capacitance/yield stress magnitude interdependence is important since the time constant R x C restricts the frequency of operation and yield stress depends on polarisation; to what extent? Is this not a fundamental question to be answered at the outset of a fluid development programme? Do the highest shear stress and fast   t* performance have to come at the highest current flow and temperature? To a large extent the operating temperature of a fluid will be fixed by its flexible duty cycle.

8. All step voltages (anti pseudo hysteresis binary excitation) include AC and DC components; what then does a good AC or DC fluid imply? Can the temperature problem be overcome by the use of MR fluids given their long time constants?

9. Are chain structural models etc. relevant to high shear rate flow? τ static— τ kinematic and seems to apply at high accelerations.

10. Are general electro fluid characteristics of paramount consideration versus lubrication properties? It has been shown that sample fluids are capable of providing the all important boundary lubrication. Particles do not seem to provide much help in areas of expected high elastohydro-dynaxmc lubrication. Tests on Rayleigh hydro dynamic bearings have shown that ER fluid behaves as a continuum generating pressures predictable from the Couette shear mode data: their pressure generation values seem inferior to Poiseuille flow and Couette flow values - why?

11. Will binary excitation overcome electro phoresis?

12. Hysteresis problems: are the situations shown in Fig. 8 due to the same cause? If so what is that cause? Is bang-bang operation necessary where hysteresis is a problem.

13. What to do about durability testing air, water and contaminant ingress and centrifuging/high shear rate effects?

14. Is there a stability criterian descibing extent of satisfactory operation?

ELECTRIC FIELD STRENGTH, TEMPERATURE AND FREQUENCY DEPENDENCY OF THE DYNAMIC PROPERTIES OF AN ELECTRO-RHEOLOGICAL FLUID

S.O. OYADIJI

Dynamics and Control Research Group, Manchester School of Engineering, University of Manchester, Manchester M13 9PL, UK

ABSTRACT

The complex shear modulus properties of an electro-rheological fluid, composed of 50 wt.% of starch in 50 wt.% of silicone oil, was determined experimentally by the application of the direct stiffness technique and an ER fluid device consisting of concentric cylinders. The electric field strength, temperature and frequency ranges of the measurements were 0.0 to 2.0 kVmm^{-1}, 0 to 60 °C and 30 to 300 Hz respectively. The results show that the shear modulus of the ER fluid decreased by a factor of up to 20 as the temperature was increased from 0 to 60 °C, whereas the shear loss factor increased from a low value of about 0.05 at 0 °C to a high value of about 1.0 at 60 °C. Conversely, as the electric field strength was increased from 0.0 to 2.0 kVmm^{-1}, the shear modulus increased whereas the loss factor decreased. However, both the shear modulus and loss factor increased in value as the excitation frequency was increased. By means of the temperature-frequency superposition principle, master curves of shear modulus and loss factor, which vary with frequency over several decades at a constant reference temperature and for two values of electric field strength, were derived from the measured data.

1. INTRODUCTION

Electro-rheological (ER) fluids are a class of 'smart' fluids whose viscosities are changed when they are placed in an electric field of varying field strength. Essentially, ER fluids are composed of a mixture of very small particles of a polarisable or conducting material and a non-conducting fluid. As the electric field strength applied across an ER fluid is increased, it becomes more viscous; at high electric field strengths, the fluid is transformed from the liquid state to a soft solid state and behaves like a solid viscoelastic material. Correspondingly, the dynamic mechanical properties of ER fluids progressively change, albeit in a non-linear fashion, as the electric field strength is increased from the value of zero.

In general, the dynamic mechanical properties of ER fluids vary with operating variables such as electric field strength, temperature and frequency. Thus, in designing ER fluid devices such as actuators, vibration and shock absorbers, dampers and mounts, clutches and brakes, it is very important to know how the dynamic mechanical properties of the ER fluid vary with these variables. These properties are usually measured by means of rotary rheometers or viscometers which consist basically of two concentric cylinders.[1-6] The shear stress, shear strain and shear strain rate of the ER fluid under test are determined from the torque or rotary motion transmitted to the inner cylinder and the excitation applied to the outer cylinder.

A different measurement procedure is employed in the present work. An ER fluid device which consists of a nest of three fixed and three movable cylinders is used. The complex axial shear stiffness of the ER fluid cylinders is determined from the force transmitted to the fixed cylinders and the input excitation applied to the movable cylinders. This stiffness is converted to the complex shear modulus of the ER fluid by the use of the appropriate geometrical factor. This measurement procedure was used to determine the variation of the complex shear modulus properties of a silicone oil-based electrorheological (ER) fluid with electric field strength, temperature and frequency. It is shown that the shear modulus of this ER fluid increases as the electric field strength and frequency increase but decreases as the temperature increases. Conversely, it is shown that the loss factor of the ER fluid decreases as the electric field strength increases but increases as the frequency and the temperature increase.

2. DIRECT STIFFNESS TEST METHOD

The direct stiffness method is an impedance measurement technique for determining the complex dynamic stiffness K^* of a sample of a resilient material. The sample is sandwiched between the top of a vibrator and a rigid termination of theoretically infinite impedance. The end of the sample which is connected to the vibrator is subjected to either sinusoidal or random displacement excitation. In the case of sinusoidal excitation, the input displacement excitation is of the form $x(t) = Xe^{i\omega t}$. The ratio of the corresponding output force $f(t) = F^*e^{i\omega t}$ to the input displacement gives the complex dynamic stiffness K^* at the excitation frequency ω. The dynamic stiffness K, phase or loss angle θ, and loss factor η are related to K^* by

$$K = |K^*| = |F^* / X|; \quad \theta = \angle K^* = \angle(F^* / X); \quad \eta = \tan\theta \tag{1}$$

The complex dynamic stiffness K^* of the sample is related to the real part K' and imaginary part K'' of K^* by

$$K^* = K' + jK'' = K'(1 + j\eta); \quad K = |K^*| = K'\sqrt{1 + \eta^2} \tag{2}$$

By multiplying the measured complex shear stiffness K^* by the geometrical factor t/A, the complex shear modulus G^* of the sample is obtained as

$$G^* = K_s^* t / A \quad \text{where} \quad G^* = G' + jG'' = G'(1 + j\eta); \quad G = |G^*| = G^*\sqrt{1 + \eta^2} \tag{3}$$

The parameters t and A are the shear thickness and cross-sectional area, respectively, of the sample parallel to the shear direction. Thus, the magnitude G and loss factor η of the complex shear modulus are obtained from the measured displacement excitation and the transmitted force by the application of eqns 1–3.

3. EXPERIMENTAL MEASUREMENT

The ER fluid device used in the experimental tests consisted of a nest of three fixed and three movable cylinders. This device, which had an inter-electrode gap of 3 mm between adjacent pairs of cylinders, was used to determine the variation of the complex shear modulus properties of a silicone oil-based electro-rheological (ER) fluid with electric field strength, temperature and frequency. The silicone oil-based ER fluid was prepared

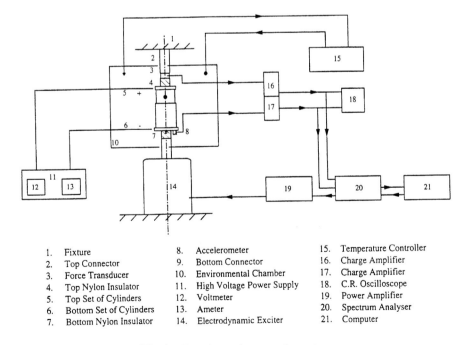

1.	Fixture	8.	Accelerometer	15.	Temperature Controller
2.	Top Connector	9.	Bottom Connector	16.	Charge Amplifier
3.	Force Transducer	10.	Environmental Chamber	17.	Charge Amplifier
4.	Top Nylon Insulator	11.	High Voltage Power Supply	18.	C.R. Oscilloscope
5.	Top Set of Cylinders	12.	Voltmeter	19.	Power Amplifier
6.	Bottom Set of Cylinders	13.	Ameter	20.	Spectrum Analyser
7.	Bottom Nylon Insulator	14.	Electrodynamic Exciter	21.	Computer

Fig. 1   Experimental test configuration.

by mixing 50 wt.% of Analar starch (analytical reagent grade) with 50 wt.% of Dow Corning silicone oil of 100 centistoke viscosity. The space between the ER fluid concentric cylinders was filled with a measured volume of the ER fluid. The ER fluid device was subsequently assembled in a direct stiffness test rig as shown in Fig. 1. This figure also shows the other items of instrumentation used for the tests and analyses of the data.

In order to control the temperature of the ER fluid, the device was enclosed within a temperature-controlled environmental chamber. The operating temperature was varied in steps from 0 to 60 °C. Temperatures below room temperature were achieved by injecting liquid nitrogen into the chamber, while temperatures above room temperature were achieved by means of electrical heating of the air within the chamber. For each temperature increment, the complex axial shear stiffness of the ER fluid device was determined firstly at an electric field potential of 0 kV and subsequently at increments of 1 kV in the electric field potential up to the maximum voltage that the power supply could deliver to the ER fluid at the particular temperature. This maximum supply voltage was observed to vary from 1 kV at 0 °C to 5 kV at 20 °C and 13 kV at 60 °C. This variation was due to the variation in the electrical and mechanical properties of the ER fluid with temperature. The movable cylinders of the ER fluid device were subjected to random vibration of 0–1000 Hz frequency bandwidth. By means of eqn 3, the complex shear modulus of the ER fluid was derived from the measured complex axial shear stiffness of the ER fluid.

4. RESULTS AND DISCUSSION

Figure 2 shows the variation of the shear modulus and loss factor of the ER fluid with applied electric potential and frequency at a constant operating temperature of 20 °C. The figure shows that when there is no applied electric field, the shear modulus has a relatively low magnitude whereas the loss factor has a relatively high magnitude. When an electric potential of 1.0 kV is applied, the shear modulus increases by a factor of about 2 while the loss factor decreases by a factor of about 2. However, further increases in the applied electric field cause less dramatic change to the values of the shear modulus and loss factor. These results indicate that there is a very significant difference in the characteristics of the ER fluid at 20 °C and for applied electric potentials of between 0 and 1 kV. Figure 2 also shows that both the shear modulus and loss factor increase in magnitude as the excitation frequency increases.

The variation of the shear modulus and loss factor of the ER fluid with applied electric field potential and frequency at a constant operating temperature of 50 °C is shown by Fig. 3. It is seen that the shear modulus increases by approximately equal amounts as the applied electric potential is increased from 0 to 5 kV. Conversely, the loss factor decreases by unequal amounts as the applied electric potential is increased over the same range. The figure also shows that the shear modulus increases in magnitude as the excitation frequency increases. However, the loss factor does not have a clear-cut variation with frequency; at some applied electric potentials, the loss factor decreases slightly

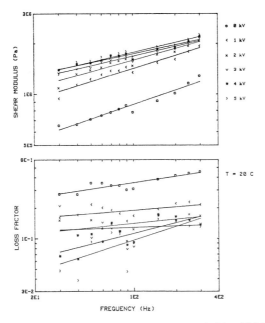

Fig. 2 Complex shear modulus properties of ER fluid at 20 °C and for applied voltages of 0 to 5 kV.

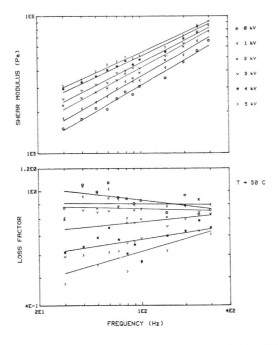

Fig. 3 Complex shear modulus properties of ER fluid at 50 °C and for applied voltages of 0 to 5 kV.

with frequency, while at other values of electric field potential the loss factor is constant or increases slightly as frequency increases. Comparing Fig. 3 with Fig. 2, it is seen that at a temperature of 50 °C, there is no sudden change in the magnitudes of the shear modulus and loss factor when the applied voltage is increased from 0 kV. The figures also show that the magnitudes of the shear modulus of the ER fluid at 50 °C are factors of between 4 and 5 less than the magnitudes of the shear modulus at 20 °C. Conversely, the loss factors of the ER fluid at 50 °C are greater in magnitude than the loss factors at 20 °C by factors of between 3 and 10.

Figure 4 shows the variation of the shear modulus and loss factor of the ER fluid with temperature and frequency at a constant applied electric potential of 1 kV. The figure shows that the shear modulus decreases whereas the loss factor increases as the temperature increases from 0 to 60 °C. Also, the shear modulus increases in magnitude as the frequency increases from 30 to 300 Hz. However, the loss factor maintains a fairly constant value over this frequency range. Similarly, Fig. 5 shows the variation of the shear modulus and loss factor of the ER fluid with temperature and frequency at a constant applied electric potential of 2 kV. By comparing Fig. 5 with Fig. 4, it is seen that the variation of the shear modulus and the loss factor of the ER fluid with temperature and frequency at 2 kV is similar to the variations observed at 1 kV. However, the magnitudes of the shear modulus at 2 kV are slightly greater than the magnitudes of the shear modulus at 1 kV, while the magnitudes of the loss factor at 2 kV are slightly less than the magnitudes of the loss factor at 1 kV. Overall, both figures show that increasing the

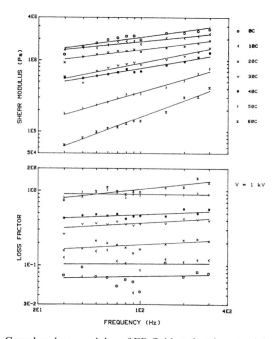

Fig. 4   Complex shear modulus of ER fluid at electric potential of 1.0 kV
and temperatures of 0 to 60 °C.

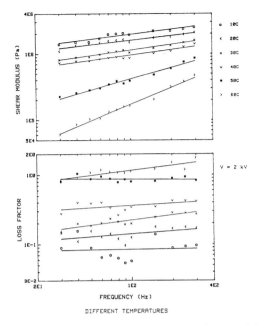

Fig. 5   Complex shear modulus of ER fluid at electric potential of 2.0 kV
and temperatures of 10 to 60 °C.

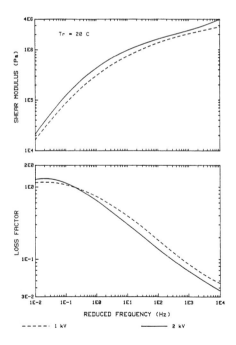

Fig. 6   Master curves of complex shear modulus of ERF at electric potentials
of 1 and 2 kV and reference temperature of 20 °C.

temperature from 0 to 60 °C has a much more pronounced effect on the complex modulus properties than increasing the excitation frequency from 30 to 300 Hz. For example, when the temperature is increased from 0 to 60 °C the magnitudes of the shear modulus are decreased by factors of up to 20 while the loss factor is increased from a mean value of 0.05 to 1.0, that is by a factor of 20. However, when the frequency is increased from 30 to 300 Hz the magnitudes of the shear modulus are increased by a maximum factor of 7 whereas the loss factor is increased by a factor of less than 2.

Using the master curve methodology, which is based on the temperature-frequency superposition principle,[7–10] the sets of frequency- and temperature-dependent complex shear modulus data shown in Figs 4 and 5 were reduced to pairs of master curves of shear modulus and loss factor at a reference temperature of 20 °C as shown in Fig. 6. It is seen that the data sets in Figs 4 and 5, which span the limited frequency range of 30 to 300 Hz and extend over the temperature range of 0 to 60 °C, have been reduced to pairs of curves which cover about 6 decades of frequency at a reference temperature of 20 °C. The similarity between the characteristics of these master curves and those of viscoelastic materials suggests that the ER fluid could be considered as an electroviscoelastic material under the conditions in which it was characterised. It can be seen from these master curves that the shear modulus of the ER fluid increases while the loss factor decreases as the applied electric field potential is increased. This is identical to the observation made earlier from Figs 4 and 5. In particular, Fig. 6 shows that when the applied electric field potential is increased from 1 to 2 kV, the shear modulus of the ER fluid is increased by about 30 % whereas the loss factor is decreased by about the same amount.

5. CONCLUSIONS

The direct stiffness method has been used to characterise the complex shear modulus properties of an ER fluid consisting of a mixture of 50 wt.% silicon oil and 50 wt.% starch. It has been shown that temperature variations affect the complex modulus properties of the ER fluid much more than variations in the excitation frequency and applied electric field potential. When the temperature was increased from 0 to 60 °C, the shear modulus decreased in magnitude by factors of up to 20 while the loss factor increased by a factor of 20. However, when the frequency was increased from 30 to 300 Hz the magnitude of the shear modulus was increased by a maximum factor of 7 whereas the loss factor increased by a factor of less than 2. By the application of the master curve method

ology, the sets of temperature- and frequency-dependent complex shear modulus data were reduced to master curves of shear modulus and loss factor which cover about six decades of the reduced frequency scale. These master curves show that when the applied electric field potential is increased from 1 to 2 kV, which correspond to electric field strengths of 0.33 to 0.67 kVmm⁻¹, the shear modulus of the ER fluid is increased by about 30 % whereas the loss factor is decreased by about the same amount.

6. REFERENCES

1. D.R. Gamota and F.E. Filisko: 'High Frequency Dynamic Mechanical Study of an Aluminosilicate Electrorheological Material', *J. Rheol.*, 1991, **35**(7), 1411–1425.
2. H. Conrad, A.F. Sprecher, Y. Choi and Y. Chen: 'The Temperature Dependence of the Electrical Properties and Strength of Electro-rheological Fluids', *J. Rheol.*, 1991, **35**(7), 1393–1410.
3. W.S. Yen and P.J. Achorn: 'A Study of the Dynamic Behaviour of an Electro-rheological Fluid', *J. Rheol.*, 1991, **35**(7), 1375–1384.
4. R. Bloodworth: 'Electrorheological Fluids based on Polyurethane Dispersions', *Proceedings of the 4th International Conference on Electrorheological Fluids*, R. Tao and G. D. Roy eds, World Scientific, Singapore, 1994.
5. D.A. Nelson and E.C. Suydam: 'The Thermal Aspects of the Electro-rheological Effect and its Impact on Application Design', *Proceedings of the Symposium on Electro-rheological Flows*, D. A. Siginer, J. H. Kim and R. A. Bajura eds, FED-vol. 164, ASME, 1993, 71–84
6. J.A. Powell: 'The Mechanical Properties of an Electrorheological Fluid under Oscillatory Dynamic Loading', *Smart Mater. Struc.*, 1993, **2**, 217–231
7. J.D. Ferry: *Viscoelastic Properties of Polymers*, John Wiley & Sons, New York, 3rd Edition, 1980.
8. A.D. Nashif, D.I.G. Jones and J.P. Henderson: *Vibration Damping*, John Wiley & Sons, New York, 1985.
9. S.O. Oyadiji and G.R. Tomlinson: 'Characterisation of the Dynamic Properties of Viscoelastic Elements by the Direct Stiffness and Master Curve Methodologies, Part 1', *J. Sound Vib.*, 1995, **188**(5), 623–647, Parts 2 & 3 (to be submitted).
10. S.O. Oyadiji: 'Characterisation of the Complex Shear Modulus Properties of Electro-Rheological Fluids by the Direct Stiffness and Master Curve Techniques', *Proc 5th Int Conf on Electro-Rheological Fluids, Magneto-Rheological Suspensions and Related Technologies*, W Bullough ed., World Scientific Publishing, Singapore, 1996; and *International Journal of Modern Physics B*, 1996, **10**(23,24), 3227–3236.

STRATEGIC ISSUES FOR THE COMMERCIAL SUCCESS OF SHAPE MEMORY ACTUATOR PRODUCTS

N.B. MORGAN AND C.M. FRIEND

*Dept. of Materials and Medical Sciences, Cranfield University, RMCS
Shrivenham, Swindon SN6 8LA*

ABSTRACT

Although recent research and development in the area of *shape memory alloys* (SMA) continues to yield novel and unique results the widespread commercial application of these materials continues to lag. This paper considers *strategic* and *commercial* issues for the application of SMA products.

Through the use of established *product management* techniques such as life cycle analysis and bipolar mapping the paper concludes that for SMA applications to become more reactive and less proactive there needs to be a pull from product design and innovation. For this to occur, the *differentiating material functions* of shape memory must be promoted and perceived as adding value to the product.

Consideration is given to the commercialisation of SMA actuators and the role of an *R&D/ Market interface*. The paper shows that for SMA actuators to become commercially viable the physical and mechanical properties of commercially available SMA's must be consolidated and future R&D focused on design properties relevant to applications.

INTRODUCTION

Driven by new technology, the materials and actuator industry is presently entering a new age. New materials and fabrication techniques are initiating a technological chain reaction that is having a knock on effect in all industries and will ultimately be crucial to their competitive advantage and market value. From medicine to aerospace new materials are being developed as solutions to specific actuation problems. This is being achieved through an overall better understanding of the fundamental principles upon which materials are based. Research and development has taken the materials industry to a point where the understanding and appreciation of the physical aspects of materials science is so high, materials may be designed on an atomic or molecular scale.

Concurrent with this revolution is the rapid globalisation of industrial business markets and a general shortening of product life cycles. The great potential of new materials such as *shape memory alloys* and the increase in global competition clearly indicates that if competitive advantage is to be achieved and maintained strategic and market issues need to be addressed.

THERMAL ACTUATORS

Thermal actuators convert thermal energy into kinetic energy. Changes in ambient or applied temperatures produce a mechanical response that results in work output. While some thermal actuators such as bimetals are well established, new actuating materials such as shape memory alloys are not. Shape memory alloys have interested those concerned with actuator design and control for some years. It is the very high work/weight ratio (work outputs can be 100 times greater than bimetals) and the inherent ability of the alloy to sense as well as actuate that is at the heart of this interest.

SHAPE MEMORY ACTUATORS

Shape memory alloys (SMA) possess unique thermal and mechanical responses. They exhibit the unusual property of a mechanical memory that may be triggered either mechanically or thermally.[1] The phenomena associated with these alloys are:

- *One-Way Memory*: where an apparently plastically deformed material will return to its original shape when heated.
- *Two-Way Memory*: where an alloy may be thermally cycled between two memorised shapes.
- *Superelasticity*: where very large and spontaneously recoverable strains may be achieved.

   The potential for these alloys to be configured into actuator applications is obvious. The use of shape memory alloys for thermal actuators offers several differentiating benefits including high forces, large strains, small size, different actuation modes (linear, bending, torsion and combinations), high work per unit volume and mass, and complete motion within a narrow temperature range.[2] The added properties in some SMA's of biocompatibility and very high recoverable strains means that the strategic importance of this class of material within the actuator and sensor industry should not be understated. Shape memory actuators may be used where large strains and high forces are required at relatively low frequencies.

   Since Greninger and Mooradian first observed thermo-elastic behaviour in the Cu–Zn system more and more alloys have been found to display shape memory effects. Of these, only Ni–Ti (NiTiNOL) and Cu based alloys (i.e. Cu–Zn–Al and Cu–Al–Ni) have proven commercially viable with useful engineering properties. Research is continuing however, into Fe based alloys that may eventually provide lower cost shape memory actuators. The Ni–Ti alloys tend to be regarded as the most useful alloys due to their superior stability, strength and corrosion resistance.

Bi-Polar Map of Functional Material Applications

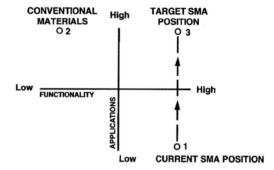

Fig. 1   Mapping of functional material applications

Figure 1 shows how these highly functional materials (position 1) lag behind conventional materials (position 2) in their applications. However, by following an appropriate strategy the applications of shape memory alloy actuators could be increased towards position 3.

LIFE CYCLE ANALYSIS

Regarding the shape memory alloy actuator as a core product allows one to explore the current stage of its life cycle. The *life cycle* shown in Fig. 2 is typical and most products follow a similar sales pattern to that shown. It is clear that although shape memory alloys have been in existence for many years they are still in the early part of the curve, i.e. at the introduction phase. Product strategy should therefore reflect this.

In the introduction phase, profits are low or even negative and competitors strategies unfocused and indirect.[3] Product strategies should aim to diffuse shape memory alloy actuators into mainstream use through those customers thought of as innovators. Careful thought and investigation, including market research, will identify who the innovators and early adopters are likely to be.

Promotion and advertising is also likely to involve heavy spending to build awareness and encourage trial amongst the innovators. This is already happening in Japan and the USA where innovative designers are well aware of SMA potential and have carried out many product trials. In the UK however the marketing of shape memory alloys has been poor and little thought has been given to appropriate promotion and placement. This has resulted in a general attitude amongst UK engineers that the use of shape memory actuators is a 'Black Art' and as such has no real commercial potential. If and when shape memory alloys are taken up by the innovators and early adopters, lower costs and higher profits will result and the product will enter the mature phase.

Life Cycle Analysis

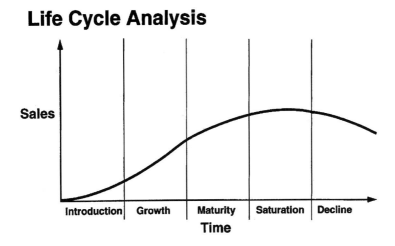

Fig. 2   Life cycle analysis

STRATEGY FOR DESIGN AND DEVELOPMENT

For shape memory alloys to succeed in real actuator applications the needs of the customer must be satisfied. This requires market research and systematic R&D on all design and product relevant properties. Failure to do this will prevent the full potential of shape memory alloys from being realised. Properties and phenomena that are inherent to the SMA such as transformation temperature, hysteresis effects, cyclic stability etc. need to be properly and reliably evaluated so as to provide the type of data essential for the design of high technology products. The development of a Ni–Ti database has already been initiated by the The Royal Insitute of Technology, Sweden[4] and a PC software program called CADSMA$_{TM}$ has been developed by A.M.T., Herk-de-Stad, Belgium and takes away the complexity of the design calculations associated with SMA actuators.[5]

To achieve cost effective and accurate property evaluation backward integration from the designer and manufacturer is necessary. Increasing alliances between universities and SMA manufacturers will increase the value and reduce the cost of research delivering the type of information essential for safe and reliable product design.[6] This type of collaboration will create an *R&D/Market interface* focusing the research onto the needs of customers.

Allied to this commercial interface is the increasing need for networks and inter-disciplinary information flow. A solid national network in the area of shape memory alloys would not only facilitate the exchange of information but attract foreign investment in what would be seen as an efficient and expert network.[6] This in turn will provide the domestic market in the UK with the type of competitive advantage required to compete with the USA and Pacific Rim countries.

PULL OR PUSH

To date shape memory based products have resulted from *technological push* based on the idea that the discovery has been made and an application needs to be found for it. Current SMA product strategies are driven by the inherent shape memory properties and these may be thought of as the *design driver*. This results in a proactive approach to applications development. Proactive strategies are typically associated with higher risk and a need for heavy and sustained investment in money and time, not only in the development and launch stages, but also throughout the product's life.[3]

Demand pull on the other hand, emphasises the fact that a customer's needs must be matched with the resources of the shape memory business. Conventional materials have tended to have their properties dictated by the product design and in this respect the design may be thought of as the *property driver*. This is a more reactive strategy and is particularly useful when markets are too small to guarantee the recovery of development costs and when only limited protection of an innovation exists. This type of market pull is beginning to occur and will intensify within the SMA market as we enter the next century. Medical and aerospace industries in particular are requiring higher performance, more 'functional' properties and the relevant developments in these areas are in turn exerting a pull on materials development. 'Nitinol' shape memory alloys may be of particular use in these applications due to their excellent work/weight ratio and corrosion resistance. However, for this strategy to work the materials designer must be responsive to the product design criteria and the design engineer must be aware of the shape memory alloy as an engineering material. In addition the *differentiating material function* of shape memory alloys must be promoted to those designers and engineers seen as innovators in their industries. The value that may be added to their products through the use of shape memory alloys will thus be apparent.

It should also be stressed to product designers that although the use of shape memory alloy actuators may increase the unit cost of the product, cost reduction is not of overriding importance in gaining competitive advantage. High value added products arising from the use of shape memory alloys will command high profit margins within a particular market segment as long as the product differentiation is of genuine value to the user/customer. When considering this strategy it is useful to look at it in the sense of a dynamic shift in price and perceived added value. Figure 3, *The Strategy Clock*,[7] shows that for a particular market segment, increasing the added value through the use of SMA's will result in a move to the top right of the diagram (A). If this strategy is to be employed it is likely that the product is aimed at a particular market segment e.g. surgical guidewires and may require a very focused approach when considering potential markets.[7]

Equally it may be that reduction in manufacturing costs through innovative design and the use of shape memory alloy actuators will actually result in no change of price or even a reduction in price, i.e. a move upwards or top left respectively in Fig. 3 (B). Again for these strategies to work the customer must be identified and the product given the appropriate added value.

The Strategy Clock

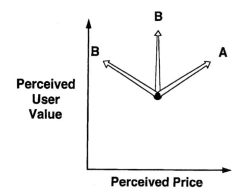

Fig. 3   The strategy clock

MARKET ENVIRONMENTS AND APPLICATIONS

Automotive manufacturing continues to dominate the industrial sector. While in the future this dominance is likely to remain, market forces will call for ever more innovative product design. Political, economic, social and technological pressures will in turn lead to a number of key factors which will be vital to maintain competitive advantage. These will include: environmental issues, safety, cost and comfort. These forces will in turn provide an opportunity for innovative shape memory actuator design and implementation. Mercedes Benz have already developed governor valves for automatic gear boxes using Ni–Ti shape memory alloys[2] and several other novel applications have been proposed using thermally activated actuators.[8]

Medical engineering is a highly opportunist market segment where shape memory alloys can make a real contribution in improving medical treatments. In recent years surgical techniques and biomedical materials have shown significant growth and the private health care industry has become a recognised market for the launch of high cost new products. Research suggests that allied to its shape memory Ni–Ti alloys have biocompatibility equal to stainless steel.[9] Current applications and research is based around endosurgery and orthopaedics.[10]

The leisure industry is another high growth/high return sector. Although the author knows of few applications or developments in this area at present, the potential is high. Either in thermal mode or super elastic configurations opportunities exist for innovative leisure products.

CONCLUSIONS

The issues addressed in this paper seek to recognise the potential of shape memory alloy actuators and their future role in products. By considering the major strategic implications of SMA products it has been shown that the diffusion of shape memory alloy materials into mainstream design usage will only come about through the successful targeting of innovative customers.

The paper recognises that the customer may only be satisfied through good and innovative design. For this to occur, accurate evaluation of commercial shape memory alloy properties must occur and the information be made available to designers. This may be achieved through close interdisciplinary collaboration and national or international networking.

It has also been proposed that the use of shape memory alloy actuators will add value to products within particular segments resulting in higher market share and increased profit margins.

REFERENCES

1. C.M.Friend: 'The Mechanisms of Shape Memory', *Proceedings of the European Symposium on Martensitic Transformation and Shape-Memory Properties*, Aussois, France, 1991, 25–34.
2. D. Stoekel: 'Shape Memory Alloys Prompt New Actuator Design', Advanced Materials and Processes, 1990.
3. R.M.S. Wilson, C. Gilligan and D.J. Pearson: *Strategic Marketing Management, planning, implementation and control*, Butterworth Heinmann Ltd, Oxford, 1992.
4. W. Tang, J. Cederstrom and R. Sandstrom: 'Property Database for the Development of Shape Memory Alloy Applications', *Proceedings of the European Symposium on Martensitic Transformation and Shape-Memory Properties*, Aussois, France, 1991, 129–134.
5. Advanced Materials and Technologies: A.M.T. n.v. Industrieweg, 43. B-3540 Herk-de-Stad, Belgium.
6. L. Kaounides: *Advanced Materials, Corporate strategies for competitive advantage*, Financial Times Management Reports, Pearson Professional Ltd, London, 1995.
7. G. Johnson and K. Scholes: *Exploring Corporate Strategy*, 3rd Ed, Prentice Hall International (UK) Ltd, 1993.
8. M.D. Perry: 'Applications Of Shape Memory Alloys In The Transportation Industry', Future Transportation Technology Conference, Society Of Automotive Engineers, Washington, USA, 1987.
9. R.S. Dutta, K. Madangopal, H.S. Gadiyar and S. Banerjee: 'Biocompatibility of Ni–Ti Shape Memory Alloy', *British Corrosion Journal*, 1993, **28**(3), 217–221.
10. F.P. Ryklina, I.Yu. Khmelevskaya, T.V. Morozova, S.D. Prokoshkin: 'Biomedical Engineering In Designing And Application Of Nitinol Stents With The Shape Memory Effect', *Proceedings Of The Third International Conference On Intelligent Materials*, Lyon, France, 1996.

APPLICATION OF MECHANICAL AMPLIFIERS IN PIEZOELECTRIC ACTUATORS

T. KING

Mechatronics Research Group, School of Textile Industries,
The University of Leeds, Leeds LS2 9JT

ABSTRACT

Piezoelectric stack or multi-layer actuators are capable of producing very large forces, but only minute displacements. Their very high rate of response and virtually zero steady-state power consumption make them potentially attractive for a wide range of machinery applications if their output displacements can be amplified to be useful in mechanisms constructed with normal engineering tolerances.

This paper discusses possibilities for mechanical amplification using flexure-hinged structures. Possible amplifier topologies are considered and appropriate techniques for the design and manufacture of flexure hinges in monolithic structures discussed.

Finally, some examples of successful amplifiers are presented including a compound amplifier structure applied to a clutching device which is capable of sub-millisecond actuation, a piezoelectric harmonic motor which utilises flexure hinged amplifiers to provide a novel type of non-magnetic stepping motor, and a low-cost roller clutch based piezomotor.

INTRODUCTION

Techniques of electro-mechanical actuation have long been dominated by electro-magnetic devices such as solenoids and rotary motors, of which a wide variety of configurations now exists. However, there has recently been increasing interest in the application of piezoelectric and electrostrictive devices,[1] particularly following the application of multilayer capacitor manufacturing techniques to the production of piezoelectric stack-type devices.[2] Moving from previous labour intensive manufacturing to the multilayer approach has led to a reduction in unit cost and significant improvements in device characteristics such as stiffness, speed, reliability, robustness and, importantly, reductions in operating voltage.[3]

Piezoelectric and electrostrictive ceramic stack-type actuators produce very small displacements but can develop large forces very quickly. By applying suitably optimised mechanical amplification techniques their output displacements can be efficiently magnified to allow their use in a wide range of applications for which electromagnetic devices are currently employed. In many cases significant advantages in terms of electrical drive speed and overall speed of response can be obtained (sub-millisecond clutching or latching is readily achieved for example). Extremely low power consumption is also possible since they require almost no power to hold a steady state.

Applications where larger movements are required necessitate a displacement amplifying linkage of some sort. Solutions to this requirement have been many and varied (eg[4]), but the simplest to manufacture are solid-state linkages based on lever principles. Devices such as simple levers where the pivots are replaced with flexure hinges (which are backlash free) can be highly efficient, but can also be bulky since highly rigid beams are needed to minimise strain energy losses in the structure.

1. DESIGN PRINCIPLES

1.1 AMPLIFIER TOPOLOGIES

Amplifying linkage topologies can be divided into two general classes; Levers and Frames.[5] Lever type devices (Fig. 1), which rely on the ratio of distances between pivots and high transverse lever stiffness to produce displacement amplification, are relatively common, although highly efficient designs are difficult to find.

Frames (quasi pin-jointed) mostly rely on the longitudinal stiffness of their constituent members rather than transverse stiffness, potentially yielding more compact designs such as that shown in Fig. 2.

Both classes rely additionally on the characteristics of flexure hinges. Conventional bearings cannot be used because of the high stresses that would be developed when the output of the amplifier becomes loaded. The use of flexure hinges also provides backlash-free rotation; an especially important characteristic in view of the extremely small displacements available directly from the actuators.

1.2 DISPLACEMENT

The movements produced by stack-type actuators can be amplified, by the use of structures such as those above, to levels of several millimetres and beyond. A material limita-

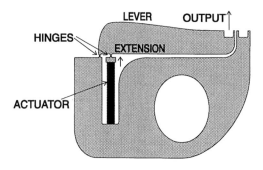

Fig. 1   Cross-section of a typical lever type displacement amplifier.

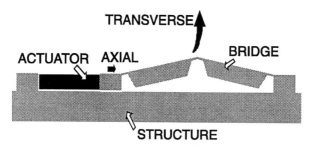

Fig. 2 Cross-section of a flexure-bridge type amplifier.

tion has to be considered however. Large displacements imply either physically large structures or large angular deflections of some of the components. The former can result in unwieldy designs which perform slowly, and the latter in high bending stresses. In most applications neither can be well tolerated but high stressing will always result in undesirably short fatigue life.

A compromise between overall size and maximum bending stress must be reached. Amplification to give several hundreds of micro-metres of output displacement has been demonstrated feasible with conventional metals such as high strength steels and titanium alloys.[6]

1.3 EFFICIENCY

The efficiency of a displacement amplifier is relevant if the output of the device is to perform work; i.e. to exert a force over a displacement. This is almost invariably the case. Since such structures behave elastically, the free output movement of the structure and the force exerted with the output stalled (fully restrained) are useful design parameters. Additionally, the mathematical product of the two relates to a work function for the structure and when compared with the same parameters for the piezoelectric actuator in isolation, provides a useful criterion of efficiency.

Modern piezoelectric multilayer stack-type actuators possess a high elastic modulus, (typically 40 GPa). When this value is compared to those for aluminium, titanium and steels, it is evident that amplifying structures made from these materials can not be considered as behaving rigidly and significant strain losses must therefore occur. Similarly, the flexure hinges dissipate energy and are a prime source of inefficiency. The loss characteristics of various metals have been studied using a finite element, semi-infinite plate model.[6,7] This has shown that when using typical commercially available actuators, the theoretical efficiency limits due to host structure stiffness limitations for Steel, Titanium and Aluminium are 81%, 71% and 49% respectively. Steel is nearly twice as stiff as Titanium but only facilitates a 10% potential improvement in force-displacement efficiency. It is almost twice as dense however, which for a device of similar proportions, would result in a slower response time.

1.4 SPEED OF RESPONSE

Electromechanical response times of typical piezoelectric multilayer actuators are in the order of 100 ms. This high-speed can not be fully exploited due to the inertia of the amplifying structure. Experimental work and theoretical modelling[7] shows that response times in the order of 1 ms can be achieved simultaneously with displacements in the order of 250 mm.

Single stage amplifiers behave approximately as second order resonant systems. A typical response to a step change in drive voltage, obtained from a piezoelectric actuator and amplifier combination, using a laser vibrometer, can be seen in Fig. 3. The ringing displayed in the lower trace can be reduced by adding a visco-elastic damper to the structure. This can be readily achieved by filling strategic points in the voids of the structure with suitably selected elastomeric compounds. It is possible to reduce the resonant response of the system to ensure critical damping, resulting in the fastest possible response, and reducing the probability of excessive 'bounce' when such structures impact onto other components (eg in clutching applications). A near-critically damped response, as shown in Fig. 4, was obtained from the same combination as for Fig. 3, after the inclusion of a quantity of polysulphide rubber into the structure.

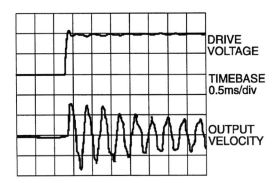

Fig. 3   Velocity response to a step voltage change (undamped).

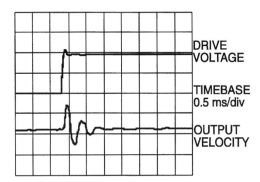

Fig. 4   Velocity response to a step voltage change (damped).

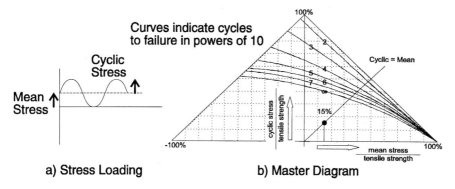

Fig. 5   Relationship between stress and cyclic fatigue.

1.5 Fatigue Life

Materials such as high-strength steels and medium strength titanium alloys exhibit the characteristic known as the stress endurance limit. Theoretically, stresses below this level should never cause fatigue.

A typical application of a flexure hinge based displacement amplifier might generate stresses within the structure similar to those shown in Fig. 5a, dependent on mechanical load and applied actuator drive voltage and waveform. Commonly, the mean stress and the cyclic stress have similar if not equal magnitude, as represented by the 45° line on the master stress diagram (Fig. 5b). Accounting for possible stress doubling caused by the structural transient response, a mean stress of 15% of the tensile strength would represent a very conservative structural stress loading which is likely to result in a theoretically limitless cyclic fatigue lifetime. Experiments have shown that the use of this kind of design regime results in structures which show no evidence of cyclic fatigue after 10^8 cycles.[8]

1.6 Temperature Stability

Mechanical amplifiers for use with piezoelectric multilayer actuators, especially those of complex design offering large displacement amplification gains, suffer from performance degradation in the form of zero position drift caused by the mismatch of the *linear coefficient of thermal expansion* of the actuator to the host structure. This can be overcome by the inclusion of a compensating element as illustrated in Fig. 6. The differential drift between actuator and host structure is governed by:

$$\left(L_p + L_c\right)\left(1 + \alpha_h \Delta t\right) = L_p\left(1 + \alpha_p \Delta t\right) + L_c\left(1 + \alpha_c \Delta t\right) \tag{1}$$

where lengths of components are L, temperature coefficients are α and suffices p, h and c refer to the piezoelectric actuator, host structure and compensator respectively. This gives:

$$L_c = L_p \frac{\left(\alpha_h - \alpha_p\right)}{\left(\alpha_c - \alpha_h\right)} \tag{2}$$

For example, in the practical case of an 18 mm long PZT actuator with a titanium host structure, the use of a zinc compensator 3.2 mm long would balance the first order thermal drift, based on the following materials characteristics.

α_p	5 x 10^{-6}	Actuator
α_h	9 x 10^{-6}	Titanium
α_c	31 x 10^{-6}	Zinc

If it is possible to closely match the elastic moduli of the structure and the temperature compensator, as in this example, the shear stresses occurring at the material boundary are minimised when the structure is loaded.

1.7 GEOMETRICAL PROFILE OF THE HINGE

Flexure hinges have traditionally been produced with a 'radial' (also termed 'right-circular') geometry, that is, they are formed from the material left in a structure after conventional drilling. Higher collective structural efficiency and reduced probability of fatigue can be achieved through the use of 'linear' or 'corner filleted' hinge designs[5] Further advantages can be obtained for particular applications by the use of more complex profiles such as elliptical profiles.[9]

Once the topology of a flexure hinged displacement amplifier is selected, the geometrical profile of the hinges will be a key factor in determining its performance.

Analytical methods were developed to design circular and right circular hinges about thirty years ago.[10] Since the *right circular hinge* is the most common profile used in flexure-hinged displacement amplifiers, the performance of this kind of hinge profile has since been further studied by both analytical and finite element methods in terms of translation precision.[11] However, little has been done by either analytical or FE methods

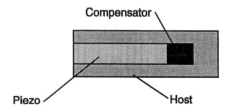

Fig. 6 Method for zero-drift compensation.

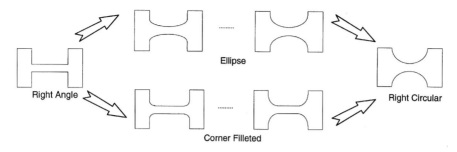

Fig. 7   Two groups of hinge profiles.

for other profiles, such as *elliptical* and *corner-filleted* designs. A finite element analysis method has therefore been employed to investigate the performance of these profiles.

1.7.1 Flexibility and Accuracy

In order to investigate the deflection and accuracy characteristics of different hinge profiles, two groups of FE model were created for analysis, one with elliptical hinge profiles (with varying minor axes) and the other with circular corner fillets (of varying radius).[12] The right angle and right circular profiles may be regarded as limiting cases of both groups as shown in Fig. 7.

The geometric configurations of the two groups of FE models investigated are:

Hinge/model number	1	2	3	4	5	6	7	8
Ellipse (minor axis mm)	0	0.25	0.5	1	2	3	4	5
Corner-filleted (radius mm)	0	-	0.5	1	2	-	4	5

All hinges have a minimum centre thickness of 1 mm, a length of 10 mm and a width of 5 mm. Steel material of E=208 GPa is assumed. Figure 8 shows a typical finite element meshed hinge model. Displacement restrictions are applied on the left edge of the model in all directions, but rotation movements are allowed. In the real structure of the displacement amplifier, a hinge will not only undergo shear stress but also a bending moment. The force applied on the edge node will produce both shear stress and bending moment in the hinge area. In the FE model, a 10 N force is applied on the centre node of the right edge.

FE analysis results are given in the form of node displacements, for the two groups of hinge profiles. The displacement of two specially defined nodes (shown as *C* and *F* in Fig. 8) are used to present two important performance characteristics of the hinges; their *flexibility* and their *accuracy*. The geometrical location of the special nodes is the same for all of the unloaded models. Node *C* is in the centre of the hinge. Node *F* is where the deflecting force is applied.

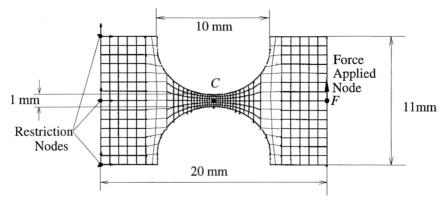

Fig. 8   A meshed FE hinge model showing boundary conditions.

An ideal hinge should be able to move freely around its central point, but a real flexure hinge requires a force to deflect it. Obviously, a hinge which needs less force has a higher flexibility. No matter how small a force is used to deflect a flexure hinge, it will offset the centre of the hinge from its geometrical centre, as shown in Fig. 9. This offset will affect the accuracy of the flexure hinge and have an effect on the displacement amplifier (of particular importance for precision oriented displacement amplifiers).

To study the flexibility quantitatively, the deflection of the node at which the force is applied (Node F) is taken to characterise the flexibility of the hinge. The offset of the centre node C is used to characterise its accuracy. The FE results of flexibility and accuracy of the two groups of hinges are shown in Figs 10 and 11. They indicate that corner-filleted hinges have a higher flexibility than the elliptical ones when the minor axis of the ellipse is equal to the corner-fillet radius, but a lower accuracy. The general trend of the two groups is similar in that the flexibility increases with decreasing ellipse minor axis and corner-fillet radius, while the accuracy decreases.

The profile of the flexure hinges affects not only the flexibility of the displacement amplifier but also the maximum stress in the hinges. Working under high stress cyclic load conditions significantly reduces the fatigue life of the hinge. Therefore a design compromise must be reached on the profile of hinge.

1.7.2 Maximum Stress

Figure 12 shows the relationship between deflection and maximum stress in the hinges under the different deflection forces (10N, 30N and 40N). It indicates that elliptic hinges are able to achieve higher flexibility with relatively lower maximum stress. The best compromise between high flexibility and low maximum stress is achieved, in this model case, with an elliptic hinge profile with a minor axis around 0.5 mm, i.e. 5% of the hinge length.

Figure 13, generated from FE results for elliptical hinge profiles (including the limiting right angle and right circular cases) shows that the location of the maximum stress point varies for different profiles. The maximum stress in a flexure hinge is produced by

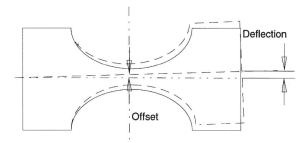

Fig. 9 Offset and deflection of a flexure hinge.

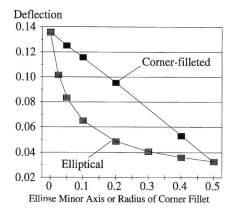

Fig. 10 Flexibility of elliptical and corner-filleted hinge shapes.

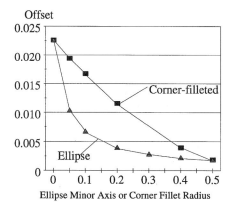

Fig. 11 Accuracy of elliptical and corner-filleted hinge shapes.

the bending moment which is used to deform the flexure hinge. Its location implies a 'bending point'. When this point is located in the centre of hinge, the flexure hinge has the highest accuracy. However the accuracy of flexure hinge is reduced when the high

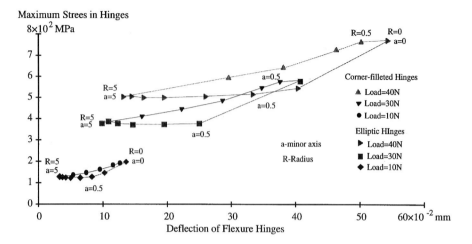

Fig. 12   Deflection of hinge vs maximum stresses.

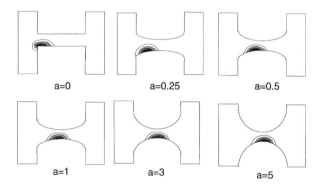

Fig. 13   Maximum stress area shift with change of length of minor axis.

stress area moves toward to the edge of the hinge from the centre. Obviously, this is because of the 'bending point' moving away from the flexure hinge centre. Figure 13 also indicates that the right angle hinge profile (a=0) concentrates stress at the corner and provides a breaking point. The right circular hinge profile concentrates stress at the centre which is the thinnest part of the hinge. the elliptical case (a=0.5) avoids both these stress concentrations.

1.8 DESIGN CONSIDERATIONS OF THE HINGE PROFILE

Depending upon their applications, flexure hinged displacement amplifiers can be divided into two groups: precision oriented and displacement oriented. A typical application of a precision oriented flexure hinged amplifier is an X–Y stage for measurement purposes,[13] whilst actuators for clutching and latching are displacement oriented applications. The design considerations of the two groups are different.

1.8.1 Precision oriented applications

Figures 10 and 11 indicate that high accuracy, unfortunately, always occurs along with low flexibility for both families of hinge profiles. However the corner-filleted hinge profile is able to achieve the highest relative accuracy when the radius of the fillet is around 5% of the hinge length. In general, for precision oriented applications which normally require displacement no larger than 0.1 mm, with a small output force, the right circular hinge profile is an acceptable one.

1.8.2 Displacement oriented applications

Amplifiers for displacement oriented application normally require not only a large output displacement but also some output force. To achieve useful output force, high efficiency is essential. The large output displacement will lead to high tensile stresses in the hinge. The design of a displacement oriented amplifier involves the selection of amplifier topology, material properties and hinge profile. Amplifier topologies and hinge profile have already been mentioned but there are two important properties of amplifier material to be considered. The Young's modulus of the material dominates the flexibility of the amplifier. The limiting tensile stress determines its fatigue life.

Use of high Young's modulus material can achieve high efficiency.[6] However, in practice, efficiency is not the only consideration in the design of monolithic displacement oriented amplifiers. The use of high Young's modulus material will inevitably lead to high tensile stresses in the flexure hinges. However, there is a limitation in using low Young's modulus material since this will reduce the stiffness of the main body of the displacement amplifier. The output force/displacement will drop significantly.

Elliptical profile hinges show better performance than corner filleted ones, especially when their minor axis length is around 5% of the hinge length, (0.5 mm in the modelled case). This is recommended as a starting point for the design of high output displacement applications.

1.9 MANUFACTURING METHODS

The most effective method of manufacture for prototype structures is the use of Wire Electric Discharge Machining (EDM). This can provide a machined surface finish of adequate quality and accuracy. It is, however, a relatively expensive method of manufacture, such that for typical amplifier structures the machining costs are much higher than the material costs, even using titanium 6/4 alloy. Costs could be reduced in mass-production by using a combination of manufacturing methods. Arrays of units could be blanked with excess material around the high-stress and tolerance critical zones. EDM could then be used to fine-machine the flexure hinges and the areas which house the actuator. Alternate schemes can be envisaged which use other methods to obtain the shaped blanks, such as laser machining.

Amplifiers for the prototype piezomotors (described below) were constructed in monolithic form by wire EDM using a Charmilles Technologies ROBOFIL 200 CNC machine. The use of a fine wire electrode (0.25 mm) provides freedom to produce a wide

Fig. 14   Hinges produced by wire EDM
(the hinge just above the centre of the picture is 0.3 mm thick).

range of hinge profiles and allows them to be machined without significantly stressing the structure; an important consideration in view of the small feature dimensions involved. Figure 14 shows an enlarged view of hinges produced by this method.

2. APPLICATION EXAMPLES

Systems currently dominated by electromagnetic devices such as solenoids and motors are obvious targets for the application of piezoelectric devices. A major advantage of piezoelectric systems over electromagnetic ones is their near-zero power requirement in situations where a static displacement or force is to be maintained for long periods. For modest displacements up to 0.5 mm, they are invariably faster and in all cases more efficient. Several examples developed by the author and co-workers are briefly described below.

Three examples of flexure hinge based amplifiers are given. All structures are manufactured from Titanium 6/4 alloy. The first two house type NLA 5x5x18 actuators (manufactured by Tokin, or equivalent devices from Physik Instrumente), which can develop a full scale movement of 15 μm and a stall force of 850 N (when displacement is fully restrained). The third design uses two such actuators stacked together to increase the available displacement. Two prototype piezomotors based on this design are briefly described.

2.1 Single-stage Actuator.

The single stage amplifier as shown in Fig. 15, is based on the principle of a simple lever and develops a full range movement of 200 μm and a stall force of 32N, representing a mechanical efficiency of approximately 50%.

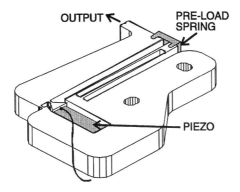

OUTPUT PRE-LOAD SPRING PIEZO

Fig. 15   Single stage amplifier – (produces 200 μm displacement).

The resonant frequency of the undamped structure is approximately 700 Hz and it can achieve its full displacement in approximately 1.5 ms. Pre-load to the piezo-stack is applied by a cantilever, machined integral with the structure, which is connected to the output arm of the actuator by a small steel clip.

2.2 MULTI-STAGE 'STRIP-CLUTCH' ACTUATOR.

Figure 16 shows an amplifier designed for rapid gripping of a reciprocating metal strip. Amplification is achieved in two stages, the first being a pair of simple levers and the final stage, a flexural bridge. The actuator displacement is magnified to develop a free movement of 0.11 mm. When stalled (at zero displacement), the output develops a force of 20 N. It can develop its full output movement within 300 μs. (Figs 3 & 4 show the response of this particular device).

2.3 PIEZOMOTORS

Piezomotors, which use piezoelectric instead of electromagnetic driving mechanisms, have been claimed to be capable of providing high torque at low speeds and to allow very precise positioning.[14] The *ultrasonic motor* was one of the first successful forms. Piezomotor designs have been reported using various drive methods.[15] A motor, proposed and built in the USSR in 1978 by Vasiliev *et al.*,[16] uses a Langevin vibrator. A travelling-wave principle has been used by AEG.[17] A *fluid coupling* ultrasonic motor has been developed by Nakamura *et al.*[18] with a rotational mechanism similar to a stationary-wave ultrasonic motor, but with force transmission by fluid coupling. The *inchworm* mechanism was originally used by Burleigh Instruments, for its patent linear piezoelectric Inchworm Motor and based on this principle, high accuracy rotational devices have been built.[19,20] Positive drives have been provided by *cycloid*[21] and *harmonic*[22] piezomotors, based on mechanical gear transmission mechanisms.

These piezomotor designs can be divided into *resonant* and *non-resonant* categories

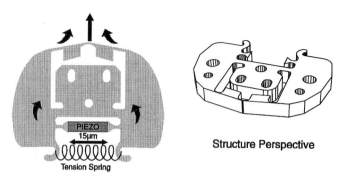

Fig. 16   Strip clutch: principle of operation and perspective view.

defined by the frequency of operation of the piezoelectric actuator. Resonant operation piezomotors have advantages of high efficiency and, in some circumstances, high output torque, however the lack of speed and positioning control ability and, for some designs, the requirement for multi-phase energisation are disadvantages. Non-resonant piezomotors generally have lower efficiency but can have other advantages, such as very precise positioning and open loop control abilities.

Non-resonant operation piezomotors, such as the harmonic piezomotor, require large displacement to be generated from the piezoelectric actuators. This is most practically achieved through the use of flexure-hinged mechanical amplifiers.

The design and construction of two different piezomotors developed by the author and co-workers is briefly described below. Although they have completely different structures and operation principles, they have similar requirements for their driving elements. A displacement amplifier has, therefore, been designed to suit both motors. A simple lever topology was selected because it uses less hinges than other topologies and therefore provides higher efficiency. In order to achieve as large an output displacement as possible, a large input displacement is required. Therefore two PI P842.10 stack-type piezoactuators, each 18 mm long and 5 x 5 mm in cross-section, were used in series. These provide a total maximum input displacement of 30 μm (at 100 V). Static output displacement achieved from the amplifier is 0.49 mm at 1N load and 0.41 mm at 2N.

2.3.1 Harmonic piezomotor

The harmonic drive principle enables the production of an externally commutated motor which can be driven as a conventional stepping motor for open-loop applications and which yields a very fine rotational increment. The use of piezoelectric actuators makes the motor power consumption almost zero in holding situations, which gives it an advantage over conventional electromagnetic motors in some applications.

The prototype harmonic piezomotor is shown in Fig. 17. It comprises displacement generating and motion converting elements. The displacement generating element is made from eight flexure-hinged displacement amplifiers. The motion converting element is constructed by adapting the gears of a harmonic drive. It converts the linear

Fig. 17 Harmonic piezomotor (partially disassembled,
with mechanical amplifier in foreground)

movement of the first part into rotational movement. Unlike the design of conventional harmonic drives where rotational wave generators are used, a 'radial spokes' wave generator is used, in which the rotational wave on the flexspline is generated by the appropriate movement of the spokes. The shape of the flexspline depends on the number of simultaneously energised piezoactuators.

The maximum required deflection of the flexure spline of the harmonic drive is equal to its tooth height plus the necessary clearance. From the configuration of the circular spline used, which has a module of 2.06, the maximum displacement requirement for the wave generator is found to be 0.41 mm.

Considering the clearance requirement, the total deformation of the flexure spline is around 0.45 mm. In view of the difficulty of manufacturing the small module gears required, the gear parts of the prototype harmonic piezomotor were re-manufactured from a production series Davall DDC-650 'Duodrive'. The thickness of the flexspline gear was reduced to 0.3 mm. The circular spline was machined to be supported by a large diameter ball-race. It works as the rotating element of the motor.

2.3.2 Characteristics of the harmonic piezomotor

For performance tests, four phase excitation at 100V p-p was provided by four MOSFET amplifiers driven by a multi-channel d-a converter installed in a PC. This configuration enabled various drive waveforms to be readily synthesized. Figure 18 shows the torque-speed characteristic of the prototype motor. (Note: the motor performs 312 'steps' per rev., so that 100 Hz corresponds to nearly 20 rev min[-1]. The maximum torques shown are analogous to the 'pull out' torques for a conventional stepping motor. Below the maximum torque limit speed is proportional to excitation frequency, as expected for a positive drive mechanism. Above the limit the spline teeth slip. More detailed descriptions of the motor and its characteristics have been published.[23,24]

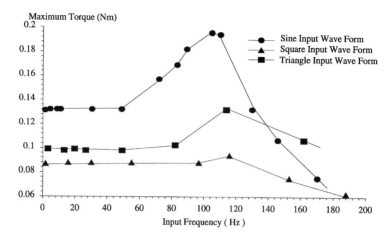

Fig. 18   Maximum torque of the prototype harmonic piezomotor.

2.3.3 Roller clutch piezomotor

The second prototype piezomotor presented is based on a roller clutch mechanism. Using one, or more, piezo-stack actuators and a roller clutch as the rotational motion converter, this motor has a very simple structure and requires only a single phase power input. To achieve effective movement of the motor using commercial tolerance roller clutches, the output movement of the piezo-stacks is again magnified by flexure-hinged displacement amplifiers. Figure 19 shows the prototype motor.

The operating principle of a roller clutch is similar to that of a ratchet mechanism. The clutch element is driven directly with circumferencially arranged piezo-actuators. The tangential push of the piezo-actuators yields rotation of the cam which is coupled to a shaft. In order to achieve rotation, two similar roller clutches are required, one to generate forward movement (driving) and the other to prevent backward rotation of the shaft when the piezo actuators return to their start positions for the next stroke (stopping). This provides stepped motion with some backlash and only unidirectional rotation. Nevertheless, relatively high speed in comparison with other piezomotors, and good efficiency are possible. The prototype motor uses a pair of displacement amplifiers driven by two piezo stack actuators. They are arranged symmetrically on opposite sides of the drive cam. The linkage between the displacement amplifier and cam is via an adjustable point contact. The driving clutch is a press fit in the cam. An HFL0408KF roller clutch is used for the driving clutch and an HF0306KF for the stopping clutch element, which is held by the frame of the motor.

2.3.4 Characteristics of the roller clutch piezomotor

To investigate performance of the motor, different input waveforms were used. The differing input waveforms affect the performance of the displacement amplifier, and hence the motor. The experimental data in Fig. 20 were measured by increasing driving frequency from 5 Hz to 150 Hz. The graph shows the characteristics of the rotational speed

Fig. 19   Prototype roller clutch piezomotor.

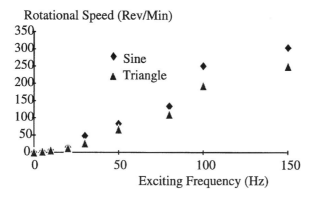

Fig. 20   Roller clutch piezomotor – no-load speed vs frequency.

against excitation frequency of the motor, in the no load condition. The results indicate that the same excitation frequency with different wave shapes yields different rotational speeds. The sine input has a slightly higher no load rotational speed than the triangle one.

Figure 21 gives the experimentally determined rotational speed vs output torque characteristic for a sine wave input. It shows that increasing the output torque reduces the output speed. This is due to the finite stiffness of the amplifier and the backlash characteristics of the roller clutches. Similar results were obtained with the triangle wave shapes. More detailed results are available.[25,26]

3. DISCUSSION AND CONCLUSIONS

Future generations of machines will require to be smaller, faster, more energy efficient and more easily computer controlled than their predecessors. This requires new actuator

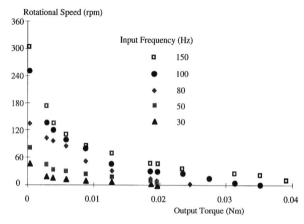

Fig. 21   Roller clutch piezomotor – speed vs torque (sine wave excitation).

designs which address the deficiencies of electromagnetic actuation in some applications. Developments in piezoceramics and piezo-actuator construction are providing higher displacement, lower voltage, devices than hitherto available. However, there is still a gap between the output displacements available from these devices and those required for a wide range of engineering applications in which practical or economic manufacturing tolerances often exceed available displacements.

The use of mechanical displacement amplifiers is a practical way of matching piezo-electric devices to machine requirements, allowing their advantages of speed and low power consumption to be realised in a range of applications including clutching, latching, pattern-selection and new types of motors.

This paper has outlined some of the considerations required when designing mono-lithic flexure-hinged amplifying structures and illustrated their successful construction and application in several areas of interest for machinery development.

As manufacturing techniques are evolved to enable economic mass production of both actuators and amplifiers their use in machinery applications will increase rapidly.

ACKNOWLEDGEMENTS

The author is indebted to Dr Wei Xu, now at the University of Dundee and Dr John Thornley, now at the University of Derby, whose PhD research work has been instrumental in developing the possibilities for mechanically amplified piezoelectric actuation. Thanks are also due to the Royal Society, who provided funds for equipment and materials used in the research described in this paper and to Davall Ltd. who generously donated the harmonic gear components modified to construct one of the piezomotors.

REFERENCES

1. T.G. King *et. al.*: 'Piezoelectric Ceramic Actuators: A Review of Machinery Applications', *Precision Engineering*, 1990, **12**(3), 131–136.
2. W. Wersing, M. Schnoller and H. Wahl: 'Monolithic Multilayer Piezoelectric Ceramics', *Ferroelectrics*, 1986, **68**, 145–156.
3. A.J. Bell: 'Piezoelectric and Electrostrictive Materials for Actuators', *Internal report*, School of Materials, University of Leeds, 1990.
4. J. Lloyd , La Comb, Jr., C.F Quate: 'Piezodriven Scanner for Cryogenic Applications', *Rev. Sci. Instrum.*, 1988, **59**(9), 1906–1910.
5. J.K. Thornley *et.al.*: 'Piezoelectric and Electrostrictive Actuators: Device Selection and Application Techniques', Proc. IMechE Eurotech Direct '91, Conf. on *Machine Systems, Drives and Actuators*, Birmingham UK, 1991, 115–119.
6. J.K. Thornley: 'Methods of Application of Piezoelectric Multilayer Actuators to High-Speed Clutching, Using Displacement Amplification', *PhD Thesis*, Loughborough University of Technology, January, 1993.
7. J.K. Thornley, T.G. King and W. Xu: 'Piezoceramic Actuators for Mechatronics Applications', Proc. *ICMA'94*, Tampere, Finland, ISBN 951-722-107-X, Feb. 1994, 569–583.
8. J.K. Thornley, M.E. Preston M.E. and T.G. King: 'A Very High-Speed Piezoelectrically Actuated Clutching Device', *Mechatronics*, 1993, **3**(3), 295–304.
9. W. Xu and T.G. King: 'Application of Flexure-Hinges to Displacement Amplifiers for Piezo-Actuators', Proc. *9th. ASPE Annual Meeting*, Cincinnati, Ohio, Oct., 1994, 258–261.
10. J.M. Paros and L. Weisbord: 'How to Design Flexure Hinges', *Machine Design*, 1965, **37**(11), 151–156.
11. S.T. Smith, D.G. Chetwynd and D.K. Bowen: 'Design and Assessment of Monolithic High Precision Translation Mechanisms', *J. Phys. E: Sci. Instrum.*, 1987, **20**, 977–983.
12. W. Xu, and T.G. King: 'Flexure Hinges for Piezo-actuator Displacement Amplifiers: Flexibility , Accuracy and Stress Considerations', *Precision Engineering* (in press).
13. F.E. Scire and E.C. Teague: 'Piezodriven 50-mm Range Stage with Subnanometer Resolution', *Rev. Sci. Instrum.*, 1978, **49**(12), 1735–1740.
14. U. Schaaf: 'Pushy Motors', *IEE Review*, 1995, May, 105–108.
15. Y. Tomikawa, T. Ogasawara and T. Takano: 'Ultrasonic Motors – Constructions/ Characteristics/Applications', *Ferroelectrics*, 1989, **91**, 163–178.
16. P.E. Vasiliev *et al.*: *UK Patent Application*, GB 2020857 A, 1978.
17. G. Schadebrodt and B. Salomon: 'The Piezo Travelling Wave Motor', *Design Engineering*, 1991, January, 36–40.
18. K. Nakamura, T. Ito, M. Kurosawa and S. Ueha: 'A Trial Construction of an Ultrasonic Motor with Fluid Coupling', *Japan. J. App. Phys.*, 1990, **29**(1), L160–L161.
19. T. Tojo and K. Sugihara: 'Piezoelectric-driven Turntable with High Positioning Ac-

curacy (first report)', *JSPE*, 1987, **53**(6), 879–884.

20. J.K. Thornley, T.G. King, and M.E. Preston: 'A Piezoelectrically-controlled Rotary Micropositioner for Applications in Surface Finish Metrology', Proc. IFToMM-jc Int. Symp. on *Theory of Machines and Mechanisms*, Nagoya, Japan, Sept. 1992.

21. I. Hayashi, N. Iwatsuki, M. Kawai, J. Shibata and T. Kitagawa: 'Development of a Piezoelectric Cycloid Motor', *Mechatronics*, 1992, **2**(5), 433–444.

22. M. Ishida, T. Hori and J. Hamaguchi: 'Principle and Operation of the New Type Motor Consisted of Piezo-electric Device and Strain Wave Gearing', *Trans. Inst. Electr. Eng. Japan D (Japan)*, 1990, **110-D**(12), 1247–1256.

23. T.G. King and W. Xu: 'The Design and Characteristics of Piezomotors Using Flexure Hinged Displacement Amplifiers', *Robotics and Autonomous Systems*, 1996, **19**, 189–197

24. W. Xu: 'Piezoelectric Motors: Aspects of Construction and Characteristics', *PhD Thesis*, The University of Birmingham, February 1996.

25. T.G. King and W Xu: 'Piezomotors using Flexure Hinged Displacement Amplifiers', *IEE Colloquium: Innovative Actuators for Mechatronic Applications*, Savoy Place, London, 18 Oct, 1995, Digest No. 1995/170, 11/1–11/5. ISSN 0963–3308.

26. W. Xu and T.G. King: 'A New Type of Piezoelectric Motor Using a Roller Clutch Mechanism', *Mechatronics*, 1996, **6**(3), 303–315.

PIEZOELECTRIC MULTI-LAYER ACTUATORS – DESIGN AND APPLICATIONS

A. DARBY

Morgan Matroc (Unilator Division) Ltd, Vauxhall Industrial Estate, Ruabon, Wrexham, Clwyd LL14 6HY

ABSTRACT

Piezoelectric multi-layer actuators, based on traditional multi-layer capacitor designs, have been designed and tested, and proved robust in service. These devices are manufactured from "soft" piezoelectrics, obtaining larger displacements than hard types, and lend themselves toward d.c. and low frequency applications. Such devices are particularly useful in fast-response applications, where large displacements are less important than operating time (e.g. if the displacement must be achieved in ~ 1 ms). Such applications include ppm bleed-valves in devices such as mass spectrometers.

Multi-layer actuators have also been fabricated from designs based around the unimorph. Applications for these devices include ink-jet printer heads, proportional valve actuation, and adaptive aerofoil technologies. Actuators have been manufactured with thickness ranging from less than 1 mm, to greater than 5 mm.

1. INTRODUCTION

Piezoelectric transducers are increasingly being used in technical applications. They are not only used as sensors to detect mechanical or acoustic impulses, but also as actuators. For actuator applications it is essential that the materials have piezoelectric d_{ij} coefficients which are as great as possible for the actuator's mode of operation. These requirements make ceramics based on the solid solution system of lead zirconate titantate (PZT) the most important group of piezoelectric materials today.

Actuators fabricated from these materials have several decisive advantages over electromagnetic actuators, including reproducible displacement, fast response times, small size, low heat generation, and do not generate electromagnetic noise. However, the driving voltages required to yield displacements usable for actuation (0.5–2 kV mm^{-1}) are too high for usual operating voltages (5–15V).

The piezoelectric d_{33} coefficient is independent of the thickness of the ceramic. Therefore, by manufacturing a stack of very thin electroded layers (<100 μm), and connecting these layers electrically in parallel, a much greater displacement in the direction perpendicular to the plane of the layers can be produced at usable voltages (<150 V).

The limiting commercial factor for production of multi-layer piezoelectric actuators is the sintering temperature of the ceramic. For most PZT compositions, the sintering temperature is beyond the usable temperature range of silver-palladium, which is the standard electrode material used in the multi-layer capacitor industry. Higher sintering temperature ceramics necessitate the use of platinum electrodes, driving up the price of the actuator. It is essential, therefore, that a PZT ceramic composition be devised for these applications with the highest possible d_{ij} coefficients, which sinters at temperatures below 1140°C.

2. DESIGN OF MULTI-LAYER ACTUATORS AT UNILATOR DIVISION

The conventional design of a multi-layer actuator is based around that of the multi-layer capacitor. Layers of thin ceramic tape are coated with a silver-palladium electrode, and stacked in an off-set arrangement, as shown in Fig.1. This configuration yields an actuator built from many smaller actuators, which are connected electrically in parallel, and mechanically in series. Therefore, the collective displacements from each layer generate a large macroscopic displacement.

Piezoelectric multi-layer actuators offer several advantages over electro-magnetic actuators. Such devices have reaction times in the micro-second range, permitting their use in a range of applications from micro-valves to active vibration damping. Furthermore, they are free from the generation of electromagnetic noise, are volumetrically and electrically efficient, and do not suffer from the problem of backlash. As these devices are solid-state elements, reliability is very good, and the screen-printing techniques employed to manufacture the actuators permits the production of designs which are limited only by the capability to dice the parts in the green state.

2.1 THICKNESS-MODE MULTI-LAYER ACTUATORS

Thickness-mode multi-layer actuators are based around the most common mode of actuation, where the driving field is applied in the direction of poling, making use of the d_{33} coefficient. The displacement of each ceramic layer is given by:

$$\Delta l = d_{33} \cdot V$$

where Δl is the displacement, and V is the applied potential difference. Therefore, in a multi-layer device, the total generated displacement is given as:

$$\Delta l = n \cdot d_{33} \cdot V$$

where n is the number of electroded layers in the actuator.

Standard designs offered by Unilator Division, in the form of a 5 mm square device, 3.5 mm thick, can yield a typical displacement of ~3 mm at 150 V, with a generated force

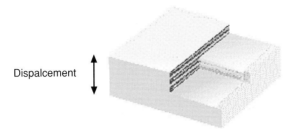

Dispalcement

Fig.1   Structure of conventional thickness-mode multi-layer actuator.

of up to 850 N. The generated force of these units is determined by the piezoelectric coefficients , the induced strain ($\Delta l/l$), the applied electric field (E/t), and the area of the electrode pattern (A). These parameters are related according to:

$$F_3 = \frac{\Delta l/A}{l S^E_{33}}$$

where S^E_{33} is the elastic compliance of the ceramic under short-circuit conditions.

It is also possible to stack individual thickness-mode multi-layer actuators. This enables the user either to increase the generated displacement for a given voltage, or reduce the applied voltage required to yield a given displacement. It is necessary, however, to consider the compliance of the bonding material used, else a component of the displacement will be lost in the bond layer.

2.2 Transverse-mode Multi-Layer Actuators

Transverse-mode multi-layer actuators rely on the d_{31} coefficient of the piezoelectric ceramic. In this type of actuator, the electric field is applied through the direction of poling, as in a thickness-mode actuator. However, the design of the device is such that they take advantage of the contraction perpendicular to the poling direction, which is connected to the expansion through the body thickness by Poisson's ratio. Such a device is depicted graphically in Fig.2.

The d_{31} coefficient is always negative when driven by an electric field with the same polarity as the direction of poling. Therefore the displacement observed in this type of actuator is also negative, and is given by:

$$\Delta l = \frac{d_{31} V l}{t}$$

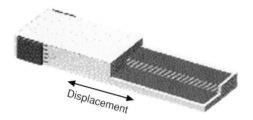

Fig. 2   Structure of transverse-mode multi-layer actuator.

where *l* is the length of the actuator, and *t* is the layer thickness.

In order to produce an actuator with a large displacement, it is necessary to optimise the aspect ratio. The greatest displacements are produced in very thin devices, but the brittle nature of ceramics results in actuators which are often too weak to be of use, else bend resulting in a reduction of the observed displacement. Transverse-mode multi-layer actuators offer the advantages of very thin layers, with a high mechanical integrity, as the displacement produced from the stacked device is equal to that of the component layers, whilst strength is gained from the increased macroscopic thickness of the device.

The contractive displacement of this type of actuator provides excellent scope for 'fail safe to closed' valve actuation. Typical actuators designed and built at Unilator Division produced displacements of up to -12 μm, with a generated force of 200 N, at 150 V.

2.3 Unimorph/Bender Multi-Layer Actuators

Conventional ceramic bimorph benders, as used in phonograph pickups, behave in a manner similar to a bi-metallic strip. Two strips of ceramic are bonded together, and may be poled in one of two ways. They can be poled in the same orientation, and then driven with opposing electrical polarity, or they may be anti-parallel poled, and driven with the same electrical polarity. These devices rely on the displacement produced due to the d_{31} coefficient, and the opposing strains derived along the length of the component causes a net bending action.

It is impractical to build multi-layer devices using this conventional method, due to the problems of electroding. However, another type of ceramic actuator exists, known as a unimorph. In these actuators, only one part of the ceramic is active. In a multi-layer actuator this may be one single electrode, or a set of electrodes that activate an area of the thickness of the part. The typical construction of a unimorph bender is shown in Fig. 3. It is worth noting that the shaded region, labelled active, consists of several electroded layers, rather than one block of active ceramic.

These actuators are idea for micro-valve actuation, and once again they can be employed as a 'fail safe to closed' type device. As is the case with conventional bimorphs, the displacement generated by multi-layer unimorph benders is greater than that of a

Fig. 3   Structure of multi-layer unimorph bender.

thickness-mode actuator, although the generated forces are much lower. Typical devices built at Unilator Division have produced a displacement of 60 mm, with a force of 24 N, at 120 V. These devices were built as prototypes for a low energy control valve[*], which could be used in a miniature robot arm. The advantages of using such a device are low power consumption, high displacement leading to small device and thus valve size, and proportional displacement. The linear nature of the relationship between displacement and applied electric field in piezoelectrics leads to displacements which are proportional to the applied electric field. Therefore, by varying the electric field, and thus displacement, it is possible to accelerate a robot arm from one position to another.

3. APPLICATIONS OF PIEZOELECTRIC MULTI-LAYER ACTUATORS

There are a great many applications which may utilise piezoelectric multi-layer actuators, which as yet have not even been considered. Current applications, however, are very diverse, ranging from valve actuation to adaptive optics. The small size, low weight, and high electrical efficiency of multi-layer actuators makes them suitable for use in the expanding field of miniaturised systems.

Valve actuation at reduced electric field strengths permits the use of these devices where, conventionally, manufacturers have avoided piezoelectric actuators. Applications include fuel injection valves, high accuracy gas analysis control valves, and inkjet printing technologies.

Piezoelectric multi-layer actuators already find use in applications such as micro-positioning, and cutting tool correction, in which tool wear is compensated for by piezoelectric actuators. In the motor industry, active damping and vibration control may be achieved by the use of multi-layer devices. The aerospace industries are considering the use of multi-layer actuators for use in SMART structures, where shape changes in aerofoils are achieved by piezoelectric actuation. In such applications, the high volumetric efficiency and low weight are of paramount importance.

The displacements produced by multi-layer piezoelectric actuators may also be employed in the optics industries. Correction of aberrations in lenses by the use of multi-layer actuators, and their use for phase correction in fibre optic transmission lines are but a few of the applications open to these devices.

[*] Designed and built for Pierre Misson, Consulting Engineer.

4. SUMMARY

The devices built and tested at Unilator Division have proven to be robust, producing usable displacements with high generated forces. Devices tested *in situ* by customers of Unilator Division, have shown excellent properties, including robust physical characteristics, good electrical properties, and have proven to be reliable and reproducible.

The shapes and modes of actuation available are only limited by the requirement to cut the devices from the tapes after the actuators have been laminated. Attempts to produce cylindrical multi-layer actuators have, so far, proved difficult, due to registration and cutting difficulties. However, once the manufacture of these devices has been achieved, a whole new group of applications are likely to emerge.

In the future, the volume of multi-layer actuator sales can only expand. As engineers become more accustomed to these devices, they will find use in an increasing number of applications. Simultaneously, the development of new ceramic compositions and manufacturing technologies will increase the range of shapes and sizes available, as well as the achievable performance.

AN INVESTIGATION INTO THE OPTIMUM DESIGN OF A PIEZO-ACTUATOR

D.F.L. JENKINS*, M.M. BAKUSH AND M.J. CUNNINGHAM

*Centre for Research in Information Storage Technology, University of Plymouth,
Drake Circusl, Plymouth PL4 8AA
Division of Electrical Engineering, Manchester School of Engineering, University
of Manchester, Oxford Road, Manchester M13 9PL

ABSTRACT

Piezoelectric elements as actuators have been extensively used to control struictures with a wide range of sizes. The effect of piezo-actuator thickness on the magnitude of deflection of a cantilever beam has been investigated experimentally and the results compared with a theoretical model. Results indicate that there exists a piezo-actuator thickness which maximises the cantilever deflection, and this optimal thickness is a function of the ratio of the Young's modulus of the actuator and the cantilever. The significance for actuator design is discussed.

1. INTRODUCTION

In recent years, piezoelectric materials have received great attention since they have proved to be effective both as electro-mechanical actuators and as sensors. If there exists an optimum thickness of piezoelectric actuator, for a given cantilever thickness, then not only can the effective bending moment be optimised, but so can the effectiveness of actuation. The actuator parameters which enable the effective bending moment to be optimised include the piezoelectric element material, its dimensions, its location and the nature of the bonding layer. The effect of the actuator thickness has been investigated theoretically by Kim and Jones[1] for plates. They produced a theoretical model for the effective bending moments produced by piezoelectric actuators bonded to stainless steel or aluminium plates. Their model showed that there is an optimum thickness of the piezoelectric actuator for maximum bending of the plate. This work investigates experimentally the effect of actuator thickness on the magnitude of the cantilever actuation. A model which describes the effective moment induced by a single piezo-actuator bonded to a cantilever beam has been presented previously.[2]

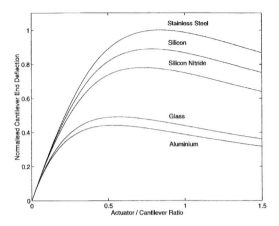

Fig. 1 Effective bending moment curves illustrating the influence of the materials.

2. THEORETICAL CONSIDERATION

For a given thickness of cantilever and actuator we have found[2] that the effective bending moment is dependent only on the Young's modulus of the cantilever, assuming the same actuator material is used. The effective bending moment for this configuration is given by:

$$M = \int_{-\frac{t_b}{2}}^{\frac{t_b}{2}} \sigma_b y \, dy = \frac{w t_b^2 \sigma}{6}$$

where w is the actuator width, t_b is cantilever thickness, σ_b is the cantilever stress and σ is the interface stress of cantilever-actuator and is given in Ref. 2.

Figure 1 shows the results obtained, and it can be seen that for each material there is an optimum thickness for maximum actuation. This result is intuitively correct since a thicker actuator produces larger actuation until the stiffening effect of the actuator itself dominates.

3. EXPERIMENTAL INVESTIGATION

In this work stainless steel and aluminium were used as cantilevers. The piezoelectric actuator used was made from PZT (Advanced Ceramics ACL4050) and bonded to the cantilever using conductive epoxy resin. The experiment was designed so that the piezo-

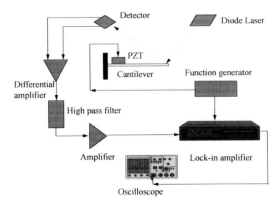

Fig. 2 Experimental arrangement.

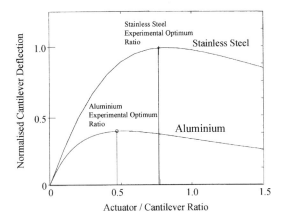

Fig. 3 Comparison of theoretical and experimental results for
aluminium and stainless steel cantilevers.

electric element thickness remained constant while the cantilever thickness was reduced.
It was very important to ensure that the cantilever could be removed from the measurement system for polishing and replaced with extreme precision. The cantilever was manually polished using carborundum powder to successively reduce the cantilever thickness. This reduction in thickness was in small incremental steps and confirmed using a
micrometer. The relative effective bending moment was determined by the magnitude of
cantilever end deflection. For this the experimental arrangement shown schematically in
Fig. 2 was used. The piezoelectric element was driven by an ac voltage of constant
amplitude, so that a constant electric field could be applied. The cantilever end deflection was measured using optical beam deflection.[3] As the magnitude of the sensed signal
is very small and accompanied by noise, the lock-in amplifier was used to recover the
signal from noise.

4. DISCUSSION

Although the model was based on certain assumptions in order to facilitate an analytical solution, Fig. 3 shows that the model prediction of an optimum actuator to cantilever thickness ratio for maximum deflection of the cantilever has been confirmed experimentally. In this figure the experimental and theoretical optimum values are determined for stainless steel and aluminium cantilevers and the agreement is good. This indicates that the theoretical model accurately predicts the optimum thickness ratio and can be used to determine the optimum actuator thickness for a given cantilever. This analysis is valid for a wide range of cantilever dimensions and materials and may be employed in many diverse applications such as micro-cantilevers used in scanning probe microscopes.

REFERENCES

1. S.J. Kim and J.D. Jones: 'Optimal design of piezo-actuators for noise and vibration control', *Am. Inst. Aeronaut. Astronaut J.*, 1991, **29**(12), 2047–2053.
2. M.J. Cunningham, D.F.L. Jenkins and M.M. Bakush: 'Experimental investigation of the optimum thickness of a piezoelectric element for cantilever actuation', *IEE Proc. Sci. Meas. Technol.*, 1997, **144**, 45–48.
3. D.F.L. Jenkins, M.J. Cunningham, W.W. Clegg and M.M. Bakush: 'Measurement of the modal shapes of inhomogenous cantilevers using optical beam deflection', *Meas. Sci. Technol.*, 1995, **6**, 160–166.

MACROCOMPOSITE TRANSDUCERS

S. DRAKE, R. LANE, AND J. CARNEY

Defence Evaluation and Research Agency, Farnborough, Hants GU14 0LX

ABSTRACT

A new sub-class of composite transducer structures described as 'macrocomposites' have been investigated by finite element modelling and practical experimentation. In current macrocomposites a ferroelectric ceramic is combined with one or more passive components on a macroscopic scale and in a regular geometric configuration to produce new electro-mechanical structures which often have significantly enhanced performance parameters compared to those possessed by their monolithic ferroelectric ceramic components in isolation. Macrocomposites can be used as sensors *and* actuators. Potential applications include miniature high sensitivity hydrophones, miniature high efficiency sound sources and high strain actuators.

1. INTRODUCTION

Piezoelectric microcomposite materials have been used for many years and can possess property combinations which are unattainable in monolithic piezoelectric ceramics. In macrocomposites one or more active ceramic components are combined with various passive components on a macroscopic scale and in a regular geometric arrangement. The stress distribution in macrocomposites is designed to make effective use of specific piezoelectric coefficients in their ceramic component(s). This allows macrocomposites to attain considerable performance increases compared with the same volume of active monolithic ceramic. The regular geometric structure of macrocomposites makes it possible to predict and optimise performance by the use of mathematical modelling techniques.

2. MACROCOMPOSITES

2.1 MACROCOMPOSITE DESIGNS

Various designs have been proposed for macrocomposite transducers. These designs aim to make maximum use of the available piezoelectric coefficients by manipulating the stress distribution in the active ceramic component(s). An important objective has been to eliminate the destructive contributions of the 3–3 and 3–1 voltage and charge coefficients when piezoelectric materials are operated in hydrostatic mode since this reduces the hydrostatic voltage and charge coefficients leading to a low hydrostatic fig-

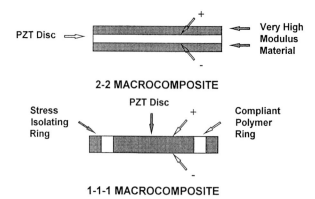

Fig. 1 2–2 and 1–1–1 macrocomposites.

ure of merit.[1] Attempts have been made in the past to restrict or isolate the 3–1 contributions as shown by the simple sandwich 2–2 macrocomposite (Fig. 1), and the 1–1–1 macrocomposite (Fig. 1) in which a PZT disc is mounted in compliant rubber within a stress isolating ring.

2.2 DISCUS MACROCOMPOSITES

A material for actuator applications was reported by Sugawara *et al.*[2] some years ago which consisted of a PZT disc with stress transforming end caps. A variety of cap shapes are possible and as these structures have been developed by various groups they have become known as 'moonies', 'discus' and 'cymbal' macrocomposites depending on the precise shape of their caps.[3]

Most work at DERA has been concerned with 'discus' macrocomposites in which the cap shape is a section through part of a hemisphere (Fig. 2). The caps of the discus macrocomposite act as mechanical transformers and change the stress distribution in the ceramic so that there is a positive contribution to the hydrostatic charge and voltage coefficients from both the 3–3 and 3–1 uniaxial coefficients (Figs 3 and 4). This results in an enhanced figure of merit. The material, shape and size of the components of the discus macrocomposite can be altered to vary the sensitivity, resonant frequency and other parameters as required by the application.

Discus macrocomposites can be used to make small high sensitivity hydrophones (Fig. 5) and compact sound sources with a low axisymmetric resonance in relation to size (Fig. 6). e.g. A F.O.M. (figure of merit) of 3×10^{-8} Pa^{-1} has been achieved in a discus macrocomposite with diameter 50 mm and an axisymmetric resonance below 4 kHz has been achieved in a discus macrocomposite with a diameter less than 90 mm.

2.3 BICERAMIC MACROCOMPOSITES

A biceramic disc consists of two piezoelectric discs bonded together so that their poling directions are opposite. When a voltage is applied across the two outer electrodes the

Piezoelectric
ceramic disc
or ring

Metal (or
plastic) caps

- — Simple structure
- — Can be used in active or passive mode
- — Very small sensitive acoustic sensors (hydrophones)
- — Compact sound sources

Fig. 2   Structure of discus macrocomposite (schematic).

For both d and g coefficients the 3-3 parameter has opposite
sign and approximately twice the magnitude of the 3-1 parameter

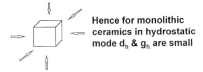

Hence for monolithic
ceramics in hydrostatic
mode d_h & g_h are small

Figure of merit for piezoelectric materials is given by $d_h g_h$ and
is thus very small for monolithic ceramics

Fig. 3   Origin of low hydrostatic figure of merit.

In discus macrocomposites the stress transforming caps
change the stress distribution in the ceramic:

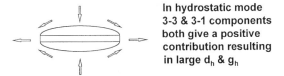

In hydrostatic mode
3-3 & 3-1 components
both give a positive
contribution resulting
in large d_h & g_h

Hence figure of merit is much increased

Fig. 4   Operation of discus macrocomposites.

diameter of one disc will increase whilst the diameter of the other disc will reduce causing the biceramic disc to bend as shown in the diagram (Fig. 7). (The same effect is
possible by bonding two discs together with their poling directions aligned and then
driving them in opposite directions by means of a central electrode.)

By using alternate ring and plug spacers, the designs of which are critical, it is possible to create a stack structure with high strain at low drive field and a low axisymmetric
resonance (Fig. 8). For example, a macrocomposite stack of six biceramic discs together

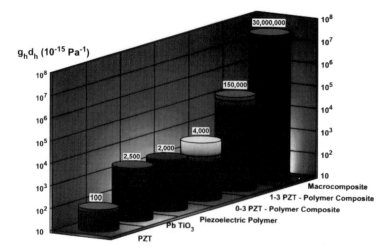

Fig. 5   Figure of merit of transducer materials.

Modelling has aided the development of discus structures with low frequency axisymmetric resonance suitable for use as underwater sound sources

e.g.

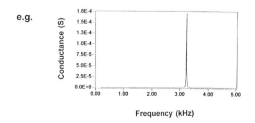

Overall diameter of structure = 90 mm

Fig. 6   Resonant behaviour of discus macrocomposite.

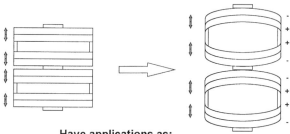

Have applications as:
• Compact sound sources
• High strain actuators

Fig. 7   Biceramic stack structure (schematic).

Fig. 8   Biceramic stack macrocomposite.

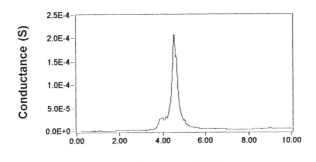

Diameter and length under 30 mm

Fig. 9   Resonant behaviour of biceramic stack.

with plug/ring spacers, in which overall diameter and length was less than 30 mm, has been constructed with an axisymmetric resonance below 5 kHz in air (Fig. 9). Complete underwater transducers have recently been constructed using biceramic stacks in which the resonance frequency in water was below 2 kHz.

3. CONCLUSIONS

The use of ferroelectric ceramics as composites enables large performance increases to be achieved. Macrocomposites have some unique attributes compared with other piezoelectric materials and will become a potential alternative to many materials presently in use.

REFERENCES

1. D. Stansfield: *Underwater Electroacoustic Transducers*, Bath University Press, Claverton Down, Bath, 1991.
2. Y. Sugawara, K. Onitsuka, S. Yoshikawa, Q.C. Xu, R.E. Newnham and K. Uchino: 'Metal-Ceramic Composite Actuators', *J. American Ceramic Society*, 1992, **75**(4), 996–998.
3. L. Cross: *Materials for Adaptive Structural Acoustic Control*, (4–6–1995) L.E. Cross ed., ONR Contract Number.: N00014-92-J-1510. Materials Research Laboratory, Pennsylvania State University, University Park, PA 16802.

TRANSMISSION ELECTRON MICROSCOPY OF THE ANTIFERROELECTRIC – FERROELECTRIC PHASE BOUNDARY IN PbZrO$_3$-BASED COMPOUNDS

I.M. REANEY* A. GLAZOUNOV[†], F. CHU[†], K.G. BROOKS[†],
C. VOISARD[†], A. BELL[§] AND N. SETTER[†]

*Department of Engineering Materials, Sir Robert Hadfield Building, Mappin St,
University of Sheffield, Sheffield, S1 3JD
[†]Laboratoire de Céramique, EPFL, MX–D Ecublens, 1015–Lausanne, Switzerland
[§]Oxley Developments Co. Ltd., Priory Park, Ulverston, Cumbria, LA12 9QG

ABSTRACT

Transmission electron microscopy and X-ray diffraction have been used to characterise the antiferroelectric (AFE) – ferroelectric phase (FE) boundary in PbZrO$_3$-based materials. Bulk ceramics in the solid solution (Pb$_{1-x}$Ba$_x$)(Zr$_{1-x}$Ti$_x$)O$_3$ (PBZT) and thin films of the general formula, Pb(Zr,Sn,Ti)$_{0.98}$Nb$_{0.02}$O$_3$ (PZSNT) were investigated. The phase changes in PBZT as a function of composition and temperature have been studied with a view to understanding the fundamental crystal chemistry across the AFE–FE phase boundary. PZSNT is a potential commercial composition which can be deposited as a thin film using sol-gel spinning. The crystallisation of thin PSZNT from the amorphous gel to the perovskite phase is discussed.

1. INTRODUCTION

PbZrO$_3$ is the best known and most frequently studied material which is known to be antiferroelectric (AFE). AFE compounds have diploles arranged in an antiparallel configuration, Fig. 1, parallel to <011> directions. The dipoles are equal in magnitude but opposite in sense and consequently, there is no net polarisation within a unit cell, nor within a domain. Ferroelectrics (FE) have dipoles arranged in a parallel configuration and a net polarisation is present across a single domain. The prototype structure of PbZrO$_3$ at high temperature is cubic but on cooling a dielectric anomaly is observed at about 230 °C. Structural studies have shown that PbZrO$_3$ has an orthorhombic unit cell at room

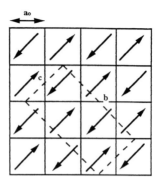

Fig. 1   Schematic representation of an antiferroelectric structure. The unit cell of PbZrO$_3$ is indicated, a = 0.82318nm (2a$_o$), 0.117764nm (2√2a$_o$) and c = O.58816nm (√2a$_o$).

temperature with a space group P2cb and lattice parameters, a = 0.82318nm, b = 0.117764nm and c = 0.58816nm.[1] However, investigations of the structure as a function of temperature using single crystals have suggested that the phase transition is more complex and a stable rhombohedral FE phase has been found to exist over a narrow temperature interval between the paraelectric (PE) and AFE phases.[1] The size of the temperature interval depends largely on the impurity concentration in the sample and, during dielectric measurements, on the applied bias field.[1] The relative ease with which the FE phase can be stabilised with respect to the AFE phase implies that the two phases are almost equal in free energy. A concept which fits the classical view of an AFE compounds.[2]

Several studies have been made of the AFE–FE phase boundaries in PbZrO$_3$-based systems. Small quantities of Ba and Sr have been shown to alter the dielectric response of the compound, reducing the temperature of the FE–AFE phase transition.[3] In addition, an intermediate, tetragonal AFE phase (AFE$_t$) has been proposed in order to explain some of the results.[3] This phase was thought to occupy a large section of the (Pb$_{1-x}$Sr$_x$)(Zr$_{1-y}$Ti$_y$)O$_3$ phase diagram close to PbZrO$_3$ but did not occur in Pb(Zr$_{1-x}$Ti$_x$)O$_3$ solid solution.[3–5] More recently, Xu et al.[6, 7] have suggested that the AFE$_t$ phase, frequently stated to be present in Sn and La–modified PZT by several authors,[3–5] was, in reality, an incommensurate phase, exhibiting several lock-in structures as a function of decreasing temperature. In effect, the anti-polar displacements had become disrupted by the presence of Sn, resulting in a long range incommensurate superlattice of polar displacements among the anti-polar. In extreme cases, when highly disordered, streaking could be observed along <110> directions in electron diffraction patterns.

For commercial use, the (Pb$_{0.97}$La$_{0.03}$)(Zr$_{1-x-y}$Ti$_x$Sn$_y$)O$_3$ solid solution system has been considered for applications such as capacitive energy storage and high strain displacement transducers.[6–9] These applications are based on the high strains and large charge storage which result from the switching of the compound from an AFE to a FE state. The purpose of the additions of Sn, Ti and La, from an applications perspective, is to broaden the temperature range over which switching can be achieved and to reduce the free energy difference between the two phases. Recently, thin films of AFE–FE phase boundary

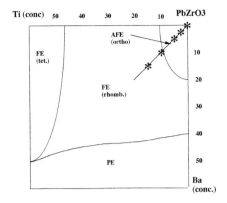

Fig. 2   PBZT phase diagram showing the compositions examined in this study.

compositions have been deposited using sol-gel spinning with promising results.[8, 9] High capacitances and strains can be achieved coupled with the potential of integration into Si-technology.

The contradictions that appear in the scientific literature demonstrate that there is incomplete understanding of the AFE–FE phase transition both as a function of temperature and composition. Dai *et al.*[10] suggested that the AFE–FE phase transition was strongly influenced by a change from antiphase rotations of O-octahedra in the AFE phase to in-phase rotations in the FE phase. The appearance of strong 1/2{hkl} reflections in the AFE phase and the appearance of equally strong 1/2{hk0} reflections in the FE were considered to be evidence of this change.[10, 11] It is the intention of the work reported here to discuss the fundamental structural changes which occur across the AFE–FE phase boundary. To investigate these changes a relatively simple, $(Pb_{1-x}Ba_x)(Zr_{1-x}Ti_x)O$ (PBZT) solid solution has been chosen. Future commercial applications may well make use of thin films deposition techniques for integration into Si-technology. Therefore, the crystallisation of sol-gel deposited $Pb(Zr,Sn,Ti)_{0.98}Nb_{0.02}O_3$ (PZSNT) has also been studied.

2. EXPERIMENTAL

PREPARATION OF SAMPLES

Pellets of the solid solution $(Pb_{1-x}Ba_x)(Zr_{1-x}Ti_x)O_3$ at x = 0, 0.02, 0.0, 0.10 and 0.15 as illustrated on on the PBZT phase diagram, Fig. 2, were prepared using Standard Laboratory Reagent (SLR) grade starting powders, PbO, $BaCO_3$, ZrO_2 and TiO_2. The powders were mixed and ball milled under acetone for 24h and calcined twice at 870 °C and 930 °C. The compound was milled after each calcination. The material was uniaxially cold pressed into a pellet and sintered at 1300 °C for 3h in a sealed crucible.

Thin films of PZSNT were prepared using the method descibed by Voisard *et al.*[8] A room temperature isothermal phase diagram for Sn doped PZT is shown in Fig. 3 illustrating the phase boundary region of interest.

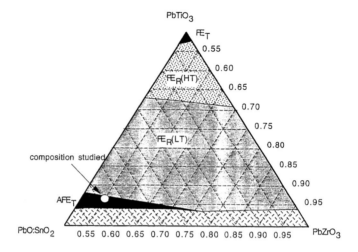

Fig. 3   Isothermal section at room temperature through
the PbO, SnO_2–TiO_2–ZrO_2 phase diagram.

DIELECTRIC MEASUREMENTS

Ceramic samples for dielectric measurements were prepared in the form of discs, 1 mm thick and 8.4 mm in diameter. Gold electrodes were deposited by sputtering. The temperature dependence of weak-field dielectric constant and loss was measured over the temperature range from -70 °C to 220 °C at 10 kHz. Ferroelecric hysteresis curves were obtained using a Sawyer–Tower bridge operating at 1 kHz. for the ceramic samples. A Radiant Technology RT66A, commercial test system was used to obtain hysteresis measurements on the PZSNT films.

ANALYSIS

An X-ray diffractometer (Kristalloflex 805, Siemens, Germany) was used with various count-times to obtain scattering diagrams. To obtain a full scan, a short count-time of 4s was used per 0.04°/2-theta.

Thin foils for TEM were prepared using a Gatan 'dual ion-mill' (model 600) operated at 5 kV with a combined gun current of 1 mA. Details of the preaparation of cross-sections are given elswhere.[12] The foils were investigated using a Philips EM430 and CM2O TEMs. A Gatan 'hot stage' was used to examine phase transitions, in situ. Hot stage experiments carried out on phase transitions in $BaTiO_3$ have suggested that the temperatures quoted in the results and discussion section are accurate within +/- 5K.

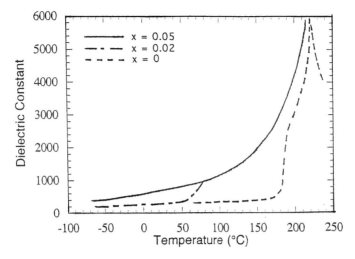

Fig. 4 Relative permittivity as a function of temperature obtained
at 10 kHz for PBZT, x = 0, 0.02 and 0.05.

3. RESULTS AND DISCUSSION

3.1 $(Pb_{1-x}Ba_x)(Zr_{1-x}Ti_x)O_3$

(a) Electrical Properties

Figure 4 shows the relative permittivity as a function of temperature (1 kHz) for 3 compositions in the solid solution $(Pb_{1-x}Ba_x)(Zr_{1-x}Ti_x)O_3$. For x = 0 ($PbZrO_3$), a peak in the relative permittivity is observed at about 225 °C. followed by a second, weaker anomaly at 185 °C. For x = 0.02, a peak in permittivity occurs at the same temperature and the second anomaly is observed at much lower temperature, 70 °C. The addition of 5%BaTiO$_3$ (x = 0.05) resulted in a single anomaly at about 225 °C; the second anomaly could not be observed for temperatures as low as -100 °C. For samples where x > 0.05, the permittivity was observed to reduce as a function of increasing BaTiO$_3$ concentration in addition to a broadening of the dielectric maxima.

Figure 5 exhibits hysteresis loops obtained at a frequency of 1 kHz at room temperature for samples where x = 0 and 0.05, respectively. The loop for x = 0 shows no evidence of either double or single hysteresis and exhibits behaviour similar to that of a linear capacitor of high dielectric loss. Samples where x = 0.02 showed identical hysteresis curves. Increasing the field resulted in electrical breakdown of the sample. This was attributed to a high level conductivity. By contrast, the loop for x = 0.05 was square, typical of that associated with a purely ferroelectric material.

(b) X-ray diffraction

Figure 6 shows X-ray diffraction spectra from samples where x = 0 and 0.05 respec-

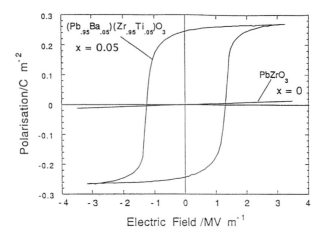

Fig. 5 Hysteresis curves obtained at 1 kHz for PBZT, x = 0, 0.05.

tively, recorded at room temperature. The spectra for x = 0 exhibits peaks which are consistent with the presence of an orthorhombic AFE phase normally observed for $PbZrO_3$ at room temperature. The same set of peaks were observed for PBZT (x = 0.02). The peaks for x = 0.05 can not be indexed according to the same orthorhombic cell but splitting of the pseudocubic {hkl} peaks (arrowed) imply that it can be indexed according to a rhombohedral cell similar to that observed for Zr-rich, $Pb(Zr,Ti)O_3$ compounds. No superlattices are present even when slow scan speeds are used to record the data. Some small peaks (arrowed) can be observed, but are attributed to unreacted ZrO_2.

(c) Transmission Electron Microscopy

Figures 7a–d are pseudocubic <100> zone axis electron diffraction patte.ns for samples where x = (a) 0 ($PbZrO_3$), (b) 0.05, (c) 0.10 and (d) 0.15. The most intense reflections in all the patterns can be indexed according to the fundamental perovskite lattice. For simplicity and direct comparison, an indexation based on this pseudocubic simple perovskite unit cell will be used at all times. Figure 7a exhibits reflections which are consistent with the presence of orthorhombic $PbZrO_3$, i.e.; strong superlattice reflections exist at the 1/4{hk0} positions. The same pattern is also found for PBZT, x = 0.02. However, Fig. 7b–d show superlattice reflections lying at the 1/2{hk0} positions and which decrease in intensity as x increases. This type of superlattice reflection can also be observed in other zone axes. It should be noted that the arcing of reflections throughout the patterns is an artefact of the preparation process. During ion-milling a thin layer of ZrO_2 crystals can sometmes form on the surface of the samples.

From a combination of the electrical data and the TEM and XRD results, it is apparent that, at room temperature, there exists an antiferroelectric orthorhombic phase for both x = 0 and 0.02. This phase is the classic AFE structure, discussed by Shirane and Hoshino[3] and more recently refined by Whatmore.[1] For x = 0.05, the XRD and electrical data

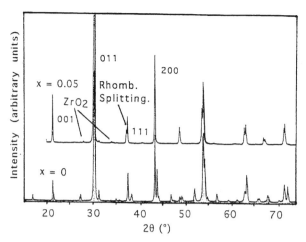

Fig. 6   X-ray diffraction spectra from PBZT, x = 0 and 0.05. Small peaks arising from unreacted ZrO_2 are indicated along with peak splitting consistent with a rhombohedral phase.

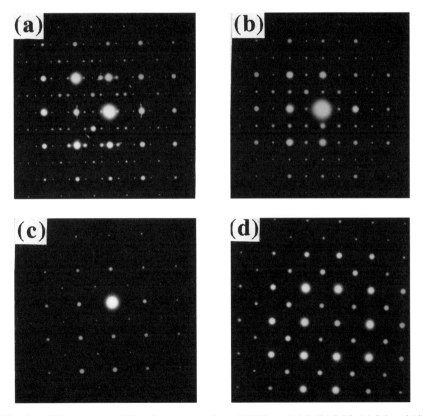

Fig. 7   <001> zone axis diffraction patterns from PBZT, x = (a) 0, (b) 0.05, (c) 0.1 and (d) 0.15. Arcing of relections is an artefact of the preparation process and is caused by thin ZrO_2 crystal that have formed on the foil surface during ion-milling.

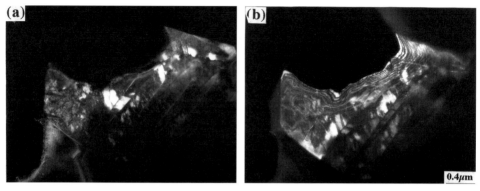

Fig. 8   Dark field TEM images obtained from a grain in PBZT, x = 0.05 using (a) 1/2(hk0) and (b) l/2(h0l) reflections and with the electron beam parallel with a <111> zone axis.

suggest that the phase is ferroelectric and similar in nature to that of the rhombohedral phase of PZT. However, the electron diffraction patterns obtained from this phase show sharp superlattice reflections at the 1/2{hk0} positions. These reflections are forbidden for the simple pseudocubic perovskite structure and a previous study[13] has suggested that these reflections arise form anti-parallel cation displacements. These displacements double the unit cell along the <110> directions.

The curious nature of the room temperature phase observed for samples where x = 0.05 (i.e.; the fact that it appears to have both parallel and anti-parallel displacements of the cations) is clearly in need of further investigation. The following dark field (DF) experiments were carried out in order to elucidate some of the microstructural features associated with this phase. Figures 8a and b are DF TEM images, from a grain in a sample where x = 0.05 obtained using 1/2(hk0) and 1/2(h0l) reflections, respectively, and parallel with a <111> zone axis, inset Fig. 8a (the contrast associated with the 1/2 (0kl) reflections was too weak to be reliably imaged). The images reveal areas of white contrast approximately 0.1 µm diameter which are asssociated with the superlattice reflections. The white contrast is highly mobile in the electron beam but it is clear that the two images are not coincident, indicating the presence of orientational domains associated with the cell doubling. In addition, macro-ferroelectric domains can also be observed. Previous studies have suggested that the doubled unit cell may be explained using the space group R3m or R3c where z = 2. The DF images however, imply that the symmetry is even lower, probably monoclimic, with the space groups, Cm and Cc as likely candidates.[13]

In order to interpret dielectric data, it is necessary to understand the structural phase transitions not only as a function of composition but also as a function of temperature. The following section will describe a series of experiments which studies in-situ the AFE/FE/PE phase transitions present within these materials. Figures 9a, b and c are electron diffraction patterns obtained fom $PbZrO_3$ with the electron beam parallel with a

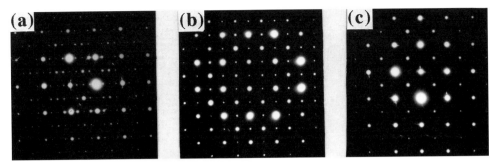

Fig. 9   Selected area electron diffraction patterns obtained from a grain of PbZrO$_3$, parallel
with a <001> zone axis and recorded at (a) 23 °C (b) 180 °C and (c) 260 °C.

<001> pseudocubic zone axis, obtained at 23 °C, 180 °C and 260 °C, respectively. Figure 9a can be indexed according to the classic AFE lead zirconate structure and has strong superlattice reflections at the 1/4{hk0} positions. At approximately 170 °C, the diffraction pattern alterred and strong superlattice reflections could only be observed at 1/2{hk0} positions, Fig. 9b. The diffraction patterns observed at this temperature were identical to those recorded for x = 0.05 at room temperature. At approximately 250 °C, the 1/2{hk0} superlattice reflections became weak and diffuse and streaking parallel with the <110> directions could be observed. Such streaks are typical of Pb-based perovskite compounds.[16] The two changes in crystal structure elucidated by the above electron diffraction patterns are coincident with the two dielectric anomalies observed as a function of temperature for PBZT, x = 0. Comparing the electron diffraction and dielectric data (Fig. 1), it becomes apparent that the first anomaly, at approximately 185 °C, can be considered as a transition from a AFE to a FE phase while the second, at about 250 °C, as a FE-PE transition. The discrepancies in the temperatures are within the additive errors for the thermocouple readings for the dielectric and hot stage measurements. The large temperature interval between the AFE and the PE-phase in this material can probably be attributed to a relatively high impurity concentration.[1]

Figures 10a, b and c are bright field images of PBZT, x = 0.02, obtained with the electron beam parallel to a pseudocubic <001> zone axis and recorded at 23 °C, 80 °C and 95 °C. At 23 °C, the grain is in the AFE phase, Fig. 10a. This phase persists until the temperature is increased to approximately 80 °C where the AFE–FE transition occurs. The onset of the phase transition is observed as the nucleation of a wedge shaped domain at the grain boundary and its subsequent growth into the centre of the grain, Fig. 10b. Figure 10c shows the same grain at 95 °C by which temperature the phase transition is complete: the domain structure is typical of that of the FE rhombohedral phase. Convergent beam microdiffraction patterns were used to monitor the phase transition by observing the changes in the nature of the superlattices within single domains.[13] The patterns confirmed the occurrence of an AFE–FE phase transition. There is good agreement

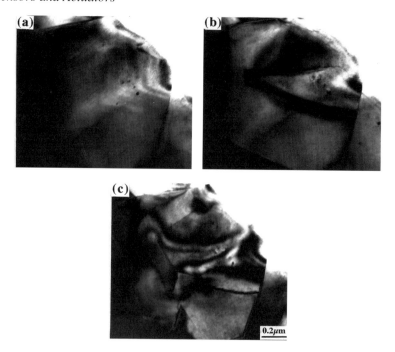

Fig. 10   Bright field TEM images showing a <001> oriented grain of PB7T,
x = 0.02 recorded at (a) 23 °C, (b) 80 °C and (c) 95 °C.

with this data and the first anomally in dielectric constant, for x = 0.02, (Fig. 1) with increasing temperature.

The results presented above show that the same phase transitions can be observed as a function of temperature and composition in PBZT. This confirms that doubling of the unit cell occurs in the ferroelectric, rhombohedral phase for samples where x = 0 as well as samples containing Ba and Ti, suggesting that the doubled ferroelectric unit cell is a general feature of the AFE–FE phase boundary.

3.2 $Pb(Zr,Sn,Ti)_{0.98}Nb_{0.02}O_3$ (PZSNT) Thin Films

(a) Dielectric Measurements

Figure 11 is a polarisation–electric field hysteresis loop from PZSNT showing a characteristic double hysteresis loop. The film was approximately 300 nm in thickness and a maximum polarization of 30 μCcm^2 was measured. Switching threshhold fields from the AFE to the FE phase as determined from relative permittivity vs dc electrical field measurements are 110 $kVcm^{-2}$ and 74 $kVcm^{-2}$ for forward and reverse switching, respectively. The dielectric constant was found to be about 460 at zero bias and room temperature and 660 at the phase transition, for measurements at 50 mV and 10 kHz.

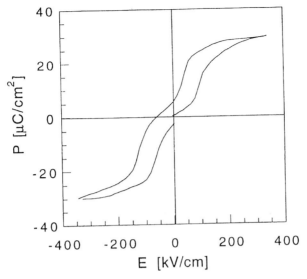

Fig. 11 Polarisation–field hysteresis curve for PZSNT showing a characteristic double loop.

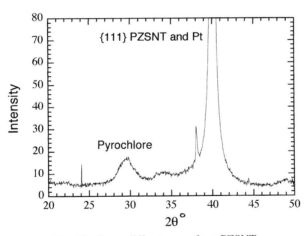

Fig. 12 X-ray diffractogram from PZSNT.

(b) X-ray diffraction

Figure 12 is an X-ray diffractogram from a thin film of PZSNT. Strong (111) orientation can be observed for the perovskite peaks. Pt and Si peaks are arrowed. A broad peak is observed at about 29° which corresponds to the pyrochlore or fluorite peak, commonly found in sol-gel derived PZT materials.[12]

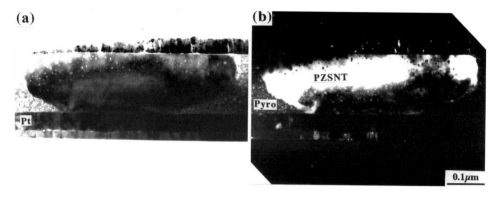

Fig. 13 (a) Bright field and (b) dark field TEM images through part of a cross section of PZSNT deposited onto Ta/Pt metallised substrates. Nucleation at the film surface is apparent.

(c) Transmision Electron Microscopy

The crystallisation of sol-gel deposited PZT has been shown to occur via the formation of a metastable pyochlore phase.[14] It is well known that lead niobate and stannate can adopt the pyrochlore structure (JCPDS, 25–443 and 17–607). Thus, it can be assumed that these additions will to some extent stabilise the pyrochlore phase, retarding the transformation to the perovskite structure. Figure 13a is a typical example of the micro-structure which is readily observed in these materials.

Although the heat teatment (750 °C for 10mm) would conventionally give sufficient energy for the transformation of pyrochlore to perovskite for pure PZT (53/47), this composition has not undergone complete transformation. Even within the apparent perovskite grains, there are particles of second phase, Fig 13b. Moreover, the in-plane grain size of the perovskite phase is of the order of 0.5–l μm compared to 0.1–0.2 μm in the pure compound, indicating a much smaller number of nucleation sites for the perovskite phase in PZSTN compared to PZT.[14] It is clear from this data that the initial premise is correct and that the addition of 'pyrochlore-formers' such as Nb and Sn will cause incomplete crystallisation of the perovskite phase. It should be noted that, in these samples, nucleation of the perovskite phase occurred at the film surface rather than the Pt surface. The adhesion layer for between Pt and SiO_2 is Ta in this case and not Ti. It is suggested that Ta may diffuse through the Pt and 'poison' the Pt surface. Ta-rich regions at the Pt/film interface would favour the pyrochlore phase preventing the fomation of stable perovskite nuclei. If the films grown on Ta/Pt metallisations are compared to those grown on Ti/Pt substrates, Fig. 14, it becomes apparent that for Ti/Pt nucleation and growth occurs on the Pt surface. It is well known that Ti diffuses rapidly through the Pt and arives at the Pt surface during crystallisation.[15] It is suggested that this allows the initial formation of a $PbTiO_3$–rich nucleus on the Pt surface allowing perovskite to grow.

Despite the promising properties of these films, the crystalline quality remains poor with a large volume fraction of second phase present. The films are therefore unsuitable for a detailed study of the AFE–FE phase boundary using in situ TEM. Recent work by

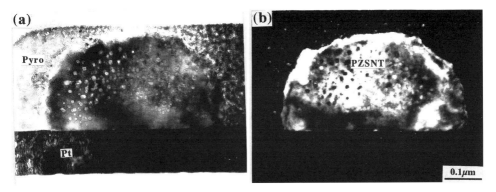

Fig. 14 (a) Bright field and (b) dark field TEM images through part of a cross section of PZSNT deposited onto Ti/Pt metallised substrates. Nucleation on Pt is observed.

Xu *et al.*[6,7] on bulk ceramics with similar compositions to the films has reported the observation of diffraction patterns much like those found in PBZT.[13] It is hoped that, in the future improved, deposition techniques will allow the complete characteristaion of the AFE–FE phase boundary in thin films.

4. CONCLUSIONS

1. The addition of up to 5% $BaTiO_3$ to $PbZrO_3$ results in the stabilisation of a ferroelectric phase over a large temperature interval.
2. The ferroelectric phase which is present in the system exhibited superlattice reflections at the $1/2\{hk0\}$ positions. A general feature of this portion of the phase diagram.
3. Although the properties of AFE - FE phase boundary thin films are promising, their crystalline quality is low, containing large amounts of second phase. The films need to be improved before further detailed study of the phase boundary is possible.

ACKNOWLEDGEMENTS

The authors would like to thank the Swiss National Foundation for Research for supporting this work. They would also like to thank the 'Centre Interdepartmentale de Microscopie Electronique' (CIME) for use their facilities.

REFERENCES

1. R.W. Whatmore: *Ph.D. Thesis*, Gonville and Cauis College, Cambridge, England, 1976.

2. M.E. Lines and A.M. Glass: *Principles and Applications of Ferroelectrics and Related Materials*, Clarendon Press, Oxford, 1977.
3. G. Shirane and S. Hoshino: *J. Phys. Soc. Japan*, 1951, **6**, 265.
4. T. Ikeda, T. Okano and M. Watanabe: *Jap. J. Appl. Phys.*, 1962, **1**, 218.
5. G. Shirane, K. Suzuki and A. Takeda: *J. Phys. Soc. Jap.*, 1952, **7**, 12.
6. Z. Xu, X.H. Dai and D. Vieghland: *Phys. Rev. B*, 1995, **51**(10), 6261.
7. Z. Xu, D. Vieghland, P. Yang and D.A. Payne: *J. Appl. Phys.*, 1993, **74**(5), 3406.
8. C. Voisard, K.G. Brooks, I.M. Reaney, L. Sagalowcz, A. Kholkin, N. Zanthopoulos and N. Setter: submitted to *Integrated Ferroelectrics*.
9. K.G. Brooks, J. Chen, K.R. Udayakumar and L.E. Cross: *J. Appl. Phys.*, 1994, **75**(3), 1699.
10. X. Dai, Z. Xu and D. Vieghland: *J. Am. Ceram. Soc.*, 1995, **78**(10), 2815.
11. A.M. Glazer: *Acta. Cryst.*, 1975, **A31,** 756.
12. I.M. Reaney K. Brooks R. Klissurska, C. Pawlaczyk and N. Setter: *J. Am. Ceram. Soc.*, 1994, **77**(5), 1209–1216.
13. I.M. Reaney, A. Glazounov, F. Chu, A. Bell and N. Setter: *Brit. Ceram. Trans.*, in press.
14. K.G. Brooks, I.M. Reaney, R. Klissurska, Y. Huang, L. Bursill and N. Setter: *J. Mater. Res.*,1994, **9**(10), 2540–2553.
15. K. Sreenivas, I.M. Reaney, T. Maeder and N. Setter: *J. Appl. Phys.*, 1994, **75**(1), 232.

ACQUISITION AND ANALYSIS OF POLARISATION-FIELD DATA FOR FERROELECTRICS

A.W. TAVERNOR

School of Materials, University of Leeds, Leeds LS2 9JT

ABSTRACT

The methods and requirements necessary to perform accurate and reproducible polarisation-field (P-E) measurements are discussed. An introduction to P-E measurements is given outlining the electrical origins of the P-E loop and visual interpretation of loop characteristics. Conventional and computer based measurement systems are reviewed followed by a description of P-E signal analysis techniques.

The DynoHYST apparatus, developed at Leeds is introduced and principles of operation are discussed. Methods and uses of electrical breakdown and ferroelectric saturation detection are presented. Particular reference is made to the dual applicability of DynoHYST to both research and production quality assurance and on line testing.

1. INTRODUCTION

Hysteresis properties of ferroelectrics have been measured for many years. The polarisation vs. field (P-E) measurement is the definitive test for many of the important material parameters and indeed for the existence of ferroelectricity itself. Conversely P-E measurements have historically been of low reproducibility. Using conventional techniques it is easy for the investigator to report false indications of ferroelectricity or unusual material behaviour. These errors are usually attributed to poor use and understanding of the apparatus. Such problems have obviously hindered the application of P-E measurements to commercial product testing and volume measurement.

Application of a large sine or triangular waveform to a linear resistor produces a P-E response as shown in Fig. 1(a). The characteristic behaviour of a linear capacitor is shown in Fig. 1(b). By comparison, a typical non-linear polar dielectric shows a saturation of polarisation events.

Ferroelectric materials possess remanence and saturation and the typical hysteresis loop shown in Fig. 1(d) results. Figure 1(e) shows a typical P-E response from a ferroelectric with poorly contacting electrodes. The gap between sample and electrode will cause a linear capacitance component to be added to the loop. This response may also be seen when measuring P-E properties of high capacitance devices, particularly multi-layers. Simple linear capacitance and resistance removal (*compensation*) may be

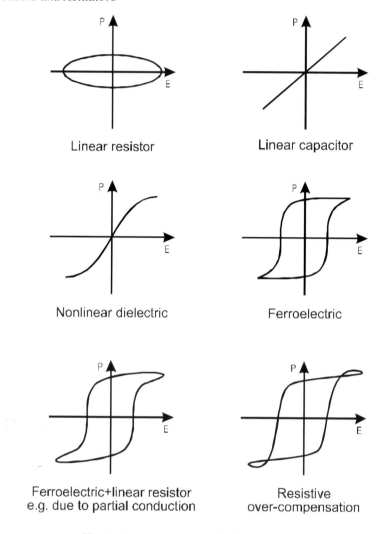

Linear resistor   Linear capacitor

Nonlinear dielectric   Ferroelectric

Ferroelectric+linear resistor
e.g. due to partial conduction   Resistive
over-compensation

Fig. 1   Typical polarisation-field characteristics

achieved via balancing circuitry. The user may adjust a variable capacitance and resist-ance until, it appears, the observation is of only non-linear behaviour. Cases of '*over-compensation*' occur frequently and result in loops characterised by Fig. 1(f). Secondary loops appear at the primary loop tips as a result of resistive over-compensation. Poor capacitive compensation results in a loop that appears not to saturate. The loop tips show a considerable gradient as a result of either resistive under- or over-compensation.

2. MEASUREMENT TECHNIQUES

2.1 SAWYER–TOWER

The most common method of performing P-E measurements is by use of a Sawyer–Tower[1] type circuit. A simple Sawyer–Tower arrangement is shown in Fig. 2.

An a.c. sine wave generator is used, via an amplifier, to apply high voltages across the device under test (DUT) and a series integrating capacitor. The integrating capacitor, C_o, is connected in parallel across the y-plates of an oscilloscope. The x-plate is connected across the a.c. voltage source. The hysteresis loop may be viewed in real time and capacitive and/or resistive compensation may be performed.

The Sawyer–Tower technique is simple and relatively cheap to perform but does possess several drawbacks. The most obvious problem is that of reproducibility. As the investigator increases the applied field, visual determination of the onset of saturation must be made. This is particularly difficult to do in most ferroelectric samples due to their high linear capacitance and resistance. Measurements are consequently made according to the *shape* of the hysteresis loop or current-time traces regardless of the precise electric field. It is possible to obtain virtually any answer with combinations of applied field strength and compensation. The problem of sample destruction is also evident. Many samples have been lost as voltage is increased excessively to ensure that saturation has been reached. This particularly occurs with samples of lower density and thicker samples where loops are less well defined and harder to saturate.

The large integrating capacitor within the circuit means that the system is somewhat restricted in frequency response but in reality this is probably no more than the frequency response limitations of an average amplifier. The integrating capacitor does, however, act to mask the non-linear components even more than the DUT capacitance. With no computerised signal analysis the technique is limited to a simple compensation.

Sine waves are conventionally used as they are less demanding on high-voltage amplifier characteristics. High-voltage triangular wave forms at useful frequencies require amplifiers with exceptional slew and switching rates. All the parameters measured by

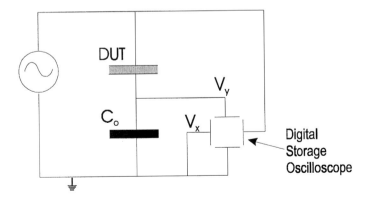

Fig. 2   Sawyer–Tower type circuit.

polarisation-field behaviour are, to lesser or greater extents, dependent on $\frac{dE}{dt}$. It is, therefore, beneficial to use a triangular wave form where the rate of change of applied field remains constant.

2.2. POST-ACQUISITION DATA PROCESSING

Computer processing of acquired polarisation-field data has been used to provide much more information than hardware only methods. The data may be obtained by digitisation of either Sawyer–Tower data or that collected directly via a current–voltage converter. The quality of the data is however critically dependent on the hardware used. Poorly acquired data will inevitably lead to inaccurate analysis. Again breakdown of samples is common and the investigator decides visually on the saturation field.

Once the data has been acquired, possibly with some form of averaging, software is used to remove unwanted components via signal deconvolution.[2–6] Analysis of the current-time data allows the calculation of all the usual P-E parameters plus many other dielectric properties. Piezoelectric, pyroelectric, resistive and capacitative properties may all be revealed by post-processing.[7]

2.3 REQUIREMENTS OF DynoHYST

The Leeds DynoHYST system has been devised in order to make P-E measurements both easier and more reproducible to make. The primary consideration was that of ensuring measurement reproducibility. For this, the decision at which field level the measurement is taken must be made via a numerical analysis of the incoming data. This requirement immediately meant that the system would utilise real-time computer control and analysis. The system would therefore be simple to use with a small amount of instruction and also capable of being used for on-line testing and production control.

The DynoHYST specification was as follows:

* Breakdown detection
* Reproducible measurement field (at a known percentage of the saturation field)
* Deconvolution of signal into linear (C and R) and non-linear components
* Measurement of true P_s and E_{sat} (i.e. not extrapolated from loop)
* Simple to use
* Modular design for maximum flexibility

The requirement for computer control also opened opportunities for the use of various types of drive fields. Within the limits of the chosen amplifier the system can generate triangle, sine, square, saw-tooth and arbitrary wave functions. The amplifier itself critically affects the type of wave that the sample experiences. An amplifier with an insufficient output current capability will radically change the wave form when driven under load at high gain. If the user requires testing using widely differing wave forms then it is probable that more than one amplifier will be required.

3. SIGNAL CHARACTERISTICS

3.1 SIGNAL DECONVOLUTION

The current-time of devices under test may be used to directly calculate polarisation-field characteristics and associated parameters. Figure 3 shows various components of current-time characteristics as a result of a triangular wave drive applied across a ferroelectric sample. The linear capacitive component (Fig. 3(b)) of the response, i_C, forms a square wave in phase with the applied field. The linear resistive component, i_R, is in phase with and proportional to the applied field (Fig. 3(c)).

The non-linear *ferroelectric* current, i_{NL}, has a form approximating to that shown in Fig. 3(d), in phase with the applied field. This current represents the rate at which the material is undergoing polarisation events (ferroelectric switching). The field at which this peak occurs, the magnitude and the symmetry of the peak together intimately describe the ferroelectric nature of the sample.

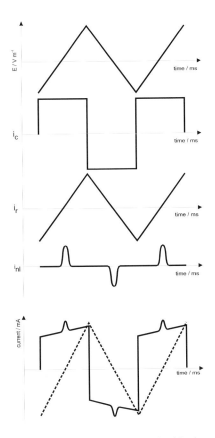

Fig. 3   Total current–time responses from a ferroelectric: (a) triangular drive field; current components of (b) capacitance, (c) resistance, (d) non-linear (ferroelectric); (e) the idealised total response of a ferroelectric under test (dotted lines show drive field).

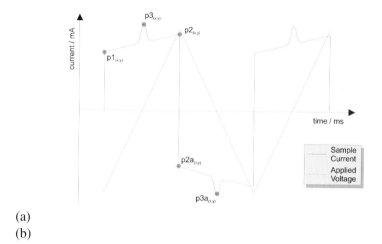

(a)

(b)

Ideal behaviour	Observed properties	Description of observation
$p2a_x = p2_x$	$p2a_x > p2_x$	Charging time
$|p1_y| = |p2a_y|$	$|p1_y| \neq |p2a_y|$	Poled ferroelectric (minor loops)
$p2_y > p1_y$	$p2_y \geq p1_y$	Linear resistive response
$(p3_x\text{-}p1_x) = (p3a_x\text{-}p2a_x)$	May be $\neq$	Poling. internal bias, other effects

Fig. 4   (a) Idealised i–t response of a ferroelectric, (b) observations from current–time responses.

Figure 3(e) shows the total (convoluted) signal resulting from both the linear and non-linear components of the sample. The relative amplitude of the non-linear current has been increased for purposes of clarity. In reality the height of this peak is considerably smaller and less discernible.

Initial investigation of the convoluted current-time response immediately reveals several ferroelectric properties. Figure 4(a). shows the convoluted i–t response. Several markers have been added to aid the descriptions, given in Fig. 4(b), of the essential features of the response.

Several of the above observations, although essentially qualitative, may be used for quality testing of production samples. Of particular interest is the degree of poling that is present in the sample – evident from various asymmetries in the response. A check could be made against a calibrated figure to ensure that production samples are properly poled.

3.2 BREAKDOWN DETECTION

Breakdown detection is the most difficult and potentially the most useful of the analyses that may be conducted using i–t data. Prior to any form of dielectric breakdown and sample damage, small increases in the linear current components are observed at the highest field levels. As the drive amplitude is increased, conduction begins to occur corresponding to the maxima and minima of the drive wave. This behaviour is depicted in idealised form in Fig. 5.

It should be noted that this effect is much easier to detect using sine wave drives rather than the more ideal triangle profile. The high field conduction starts in a much more controlled manner, using sine wave drives, and is more likely to be detected via analysis well before any sample damage. The software techniques employed do, however, allow detection of breakdown under virtually any drive conditions. Various tolerance levels may be set to allow either *fine* or *coarse* detection of breakdown.

A second indicator of the onset of breakdown is that a reduction in the magnitude of the non-linear peak may be detected. We have observed samples that have lost all non-linear behaviour after high field measurement while retaining their original linear capacitance and resistance's. This behaviour indicates a loss of ferroelectric properties but without dielectric breakdown of the sample. Without the DynoHYST software controlling the field this measurement would have probably led to total breakdown of the samples.

3.3 SATURATION DETECTION

The onset of saturation is detected in real time as the measurement field is increased. In terms of Fig. 4, and assuming an increasing amplitude of drive field then: '*Saturation is reached at the applied field level above which the linear resistive component* $(p2_y - p1_y)$ *increases proportionally to the applied field but the magnitude of the non-linear current, X, remains constant*'. This situation is represented by the two curves in Fig. 6. At this

Fig. 5   Typical i–t response with the onset of breakdown shown as dotted lines.

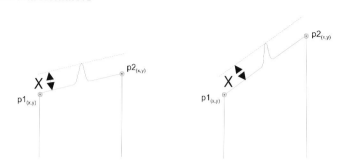

Fig. 6   Principles of saturation detection.

exact point, all the polarisation events that are possible have been undergone and increasing the field produces purely ohmic changes. Caution should be exercised in this field regime if over-riding the software control as the onset of breakdown, and consequent alteration to ferroelectric properties, is imminent.

The field level at which saturation is detected may then be used to take calibrated measurements with no fear of breakdown. Alternatively, the field may be increased until the first signs of breakdown are detected and then reduced by a known percentage to take measurements. Regardless of the method chosen, the system ensures that all measurements will be reproducible.

4.0 THE DYNOHYST APPARATUS

The DynoHYST system employs complete computer control and data acquisition to allow controlled measurement under extreme electrical stressing. The system has been designed to be modular to account for as many measurement configurations as is practicable without major system redesign, see Fig. 7.

The computer is fitted with a high performance data acquisition card with facilities for both digital and analogue input and output. The a.c. function to be applied to the sample may be either generated within the computer or via GP-IB control of an arbitrary function generator.

In order to maximise measurement resolution within a 12- or 16-bit digitisation regime the current to voltage converter (CVC) is fitted with variable output stage amplifier/attenuation circuitry. The signal coming into the computer is scaleable by both this variable stage and also the variable gain levels present on the card. The variable stage CVC is controlled via digital output lines from the computer.

Acquisition of the sample and source voltage waveforms must be conducted with care. Even when using high specification acquisition cards appreciable cross talk can be detected between channels. In our investigation, each channel combination was tried until the lowest cross talk condition was satisfied. The design of each individual card will dictate the best channels to use.

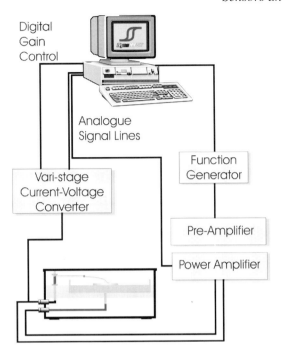

Fig. 7 Schematic diagram of DynoHYST apparatus.

The sample as shown in Fig. 8, cell comprises a machined Au-coated brass disc. The disc is machined to leave protruding walls allowing it to be filled with paraffin or other types of oil. This is useful at high fields to prevent air breakdown when it is not practicable to manufacture specialist edge defined electrode samples. One electrode connects to the disc that it is sitting on and the other is contacted using a Au-coated articulated arm assembly. The arm may be placed to contact at any point within the cell. It may also be placed in a convenient position if fibre optic displacement measurements are to be made on the sample surface. The cell is double insulated and sits within a nylon coated aluminium container acting as a Faraday cage. The whole cell may be placed in an environmental chamber and operated over the temperature range $-160 \leq T \leq +250$ °C.

5. SUMMARY

The DynoHYST measurement technique may be used to provide all the usual hysteresis loop data such as P_s, P_r, E_c and loop area. The unique DynoHYST software employed provides detection of saturation, detection of poling and detection of the onset of breakdown. This occurs in real time and is used by the software to control the field level at which parameter measurements are made. Using the acquired current-time data, calculations may be made of the capacitative, resistive, piezoelectric and pyroelectric properties.

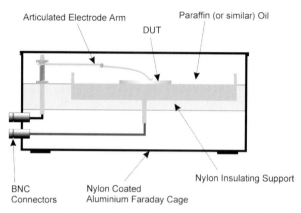

Fig. 8   Schematic of the DynoHYST sample cell.

Since the total measurement time for each sample is only approximately 10–15 seconds the DynoHYST apparatus may find application in industrial quality assurance and testing environments. In addition, the unique ability to detect saturation and breakdown behaviour may be used to determine accurate poling conditions for each batch of samples produced. The ability to detect incipient breakdown currents non-destructively also means that forensic analysis of problematic parts may be carried out (e.g. SEM) on identified devices before catastrophic breakdown has occurred. More accurate predictions of the microstructural and processing defects responsible for breakdown may then be made, rather than relying on second guessing what may have been present to cause the breakdown from a part containing burnt and melted ceramic. This technique has found particular use for forensic analysis of multi-layer components.

The ageing behaviour of devices may be investigated by monitoring the change in any of the parameters with respect to time, temperature, atmosphere or bias field. Programmable ageing functions may be applied to the DUT to simulate any *in service* degradation conditions. The sample cell has been designed to allow simple fitting of fibre optic displacement probes. This allows the comparison of piezoelectric (or electrostrictive) strain data calculated from strain field or current-time measurements on the same sample at the same time.

ACKNOWLEDGEMENTS

The author wishes to express his thanks to the DTI Materials Testing and Standards Committee and the National Physical Laboratory who provided part-funding of the hardware development under project AM4. Thanks are also due to Dr J.A. Close and Mr R. Holt for valuable discussions during development and to Dr J.A. Close for the sample holder and associated hardware design and construction.

REFERENCES

1. C.B. Sawyer and C.H. Tower: *Physical Review*, 1930, **35**.
2. M. Daglish: PhD Thesis, University of Leeds, Leeds, UK, 1990.
3. A.W. Tavernor: *PhD Thesis*, University of Leeds, Leeds, UK, 1992.
4. V.R. Yarberry and I.J. Fritz: *Rev. Sci. Instrum.*, 1979, **50**(5), 595.
5. T. Fukami, H. Yanagisawa and H. Tsuchiya: *Rev. Sci. Instrum.*, 1983, **54**(11), 1351.
6. H. Zhongliang, Y. Xi and M Zhongyan: *Proc. 6ᵗʰ IEEE Intl. Symp. of Appl. Ferroelectrics*, 1986, 726.
7. R.J. Von Rensburg, V Humberstone, J.A. Close, A.W. Tavernor, R. Stevens, R.D. Greenough, K.P. O'Connor and M.G. Gee: *Review of Constitutive Description and Measurement Methods for Piezoelectric, Electrostrictive and Magnetostrictive Materials*, DMM(A) **143**, The National Physical Laboratory, Crown Copyright. ISSN 0959 2423, 1994.

AGEING CHARACTERISTICS OF BaTiO₃ BASED PIEZOCERAMICS

D.A. HALL AND M.M. BEN-OMRAN

Materials Science Centre, University of Manchester/UMIST, Grosvenor Street, Manchester M1 7HS

ABSTRACT

The ageing and field-forced deageing characteristics of a commercial cobalt-doped BaTiO₃ piezoceramic are reported. It was found that the internal bias field in unpoled specimens reduced according to an exponential time law during field-forced deageing. The time constants associated with the deageing process were 159, 33 and 25 s at 30, 60 and 90 °C respectively, indicating an activation energy of 0.36 eV. In contrast, field-forced deageing of poled specimens yielded a logarithmic behaviour. Furthermore, the internal bias field in poled specimens proved to be extremely resistant to the deageing procedure, with the result that a significant internal bias field (E_i ≈ 0.1 kV mm⁻¹) remained after the application of a continuous AC field of 2 kV mm⁻¹ for 12 000 s at 30 and 60 °C.

1. INTRODUCTION

Barium titanate, BaTiO₃, was the first known ceramic ferroelectric and hence the first piezoelectric ceramic material. Following its discovery in 1945, much research was carried out to establish an understanding of the dielectric and piezoelectric properties of barium titanate-based ceramics and single crystals.[1]

Barium titanate ceramic dielectrics are now widely used for various types of high permittivity capacitors.[2] However, in many piezoelectric applications barium titanate has been superceded by PZT (lead zirconate titanate) ceramics, which offer better temperature stability and improved piezoelectric coefficients.[3] Nevertheless, piezoelectric barium titanate still finds use in a significant number of areas, for example in echo sounders, fish finders and air ranging transducers. For these applications, the advantages of barium titanate over PZT are related to its long-established use and economic considerations.[4]

Acceptor-doped ferroelectric ceramics exhibit pronounced ageing of dielectric properties as a function of time. As a result of this process, the dielectric permittivity and loss are reduced significantly after ageing. The reduction in loss is particularly important in allowing the production of low loss, or 'hard', piezoelectric ceramics.[3]

High field P–E measurements indicate the development of an 'internal bias field' E_i in acceptor-doped ferroelectrics associated with a gradual stabilisation of the ferroelectric domain structure. This effect gives rise to a constriction of the P–E loop for an unpoled specimen or a 'shift' of the loop along the electric field axis for a poled specimen.[5] The following empirical time laws have been observed to describe the build up of the internal bias field during ageing and its reduction during field-forced deageing:[6]

Ageing (logarithmic) : $\quad E_i(t) = A Log(t) + B$ $\hfill$ (1)

Field-forced deageing (exponential) : $\quad E_i(t) = E_i(0)e^{-\frac{t}{\tau}}$ $\hfill$ (2)

The ageing processes in acceptor-doped ferroelectrics are attributed to the presence of oxygen vacancies, which can form defect associates with acceptor ions (e.g. Ni, Cr, Fe, Co substituted for Ti).[5] The gradual re-orientation of these 'defect dipoles' towards the direction of the local domain polarisation is then held responsible for the observed domain wall stabilisation effects.[7,8]

The aim of the present work was to characterise the high field ageing and field-driven deageing behaviour of a commercial cobalt-doped $BaTiO_3$ piezoelectric ceramic. The results obtained can be compared with quantitative models of the ferroelectric domain wall stabilisation processes (as indicated above) and with similar results found previously for hard PZT materials.

2. EXPERIMENTAL PROCEDURES

A commercial cobalt-doped $BaTiO_3$ powder, type PC3, was obtained from Morgan Matroc (Unilator Division, Ruabon, UK). Green pellets were pressed at 100 MPa and then sintered at 1400 °C. According to the manufacturer's data sheet,[4] materials fabricated and poled under standard conditions typically exhibit the following dielectric and piezoelectric properties :

Table 1 Typical dielectric and piezoelectric coefficients for PC3 ceramics.[4]

ε_r	tan δ	d_{33}/pC N^{-1}	d_{31}/pC N^{-1}	k_t	k_p
1100	0.006	127	-42	0.31	0.26

Low field dielectric measurements were carried out using a Hewlett-Packard HP4284A LCR meter in combination with a wire-wound alumina tube furnace. Contacts to the specimens were made using a silver electrode paste (DuPont type 7095) fired on at 550 °C. High field P–E measurements were carried out in a temperature-controlled silicone oil bath. An applied triangular waveform with an electric field amplitude of 2 kV mm^{-1} was produced using a Thurlby Thandar TG1304 function generator together with a Chevin Research HVA1B high voltage (x1000) amplifier. The polarisation was obtained by inte-

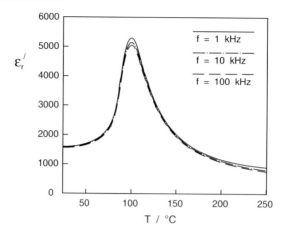

Fig. 1   Temperature dependence of permittivity for commercial cobalt-doped BaTiO$_3$ ceramics.

gration of the induced current, measured using a custom-built I–V converter. Further details of the measurement system are given in a previous publication.[9]

3. RESULTS AND DISCUSSION

Low field dielectric measurements indicated that the Curie point was around 100 °C, as shown in Fig. 1. The relatively broad peak in permittivity at the Curie point indicates a diffuse ferroelectric phase transition, perhaps due to compositional heterogeneity. A slight frequency dependence of the permittivity was also evident, but there was no indication of any change in the Curie point with measurement frequency.

The overall test procedure used for the high field ageing and deageing measurements is illustrated in Fig. 2. By following this procedure, it was possible to obtain data showing the changes in the P–E characteristics during field-forced deageing of the unpoled (aged) material, ageing of the poled material, and subsequent field-forced deageing of the poled (aged) material at various temperatures.

The initial P–E loops obtained for the unpoled, aged specimens showed a characteristic constriction at each of the ageing temperatures investigated (30, 60 and 90 °C), as illustrated in Fig. 3(a). The corresponding I–E (current–field) curves showed a pronounced 'splitting' of the ferroelectric switching peak, enabling the internal bias field E_i to be determined as half of the peak separation on the electric field axis. The value of E_i reduced rapidly during field-forced deageing, while both the remanent and saturation polarisation (P_r and P_s respectively) increased significantly (Figs 3 to 5).

The rate of reduction of E_i during deageing gave a good fit to the exponential time law (eqn 2), as illustrated in Fig. 4. The associated time constant τ values were measured as 159, 33 and 25 s at 30, 60 and 90 °C respectively. These values indicate an activation

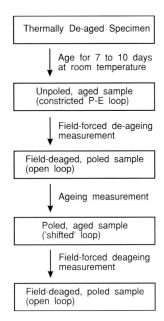

Fig. 2   Measurement procedure for high field ageing and deageing tests.

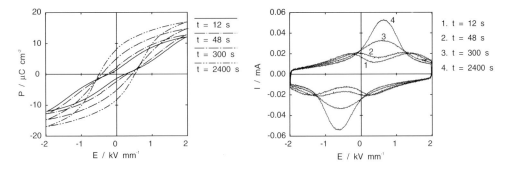

Fig. 3   Changes in (a) P–E and (b) I–E loops of unpoled materials during field-driven deageing at 30 °C (E_0 = 2 kV mm^{-1}).

energy W of 0.36 eV, assuming a temperature activated process according to eqn 3:

$$\tau(T) = \tau_\infty e^{\frac{W}{kT}} \tag{3}$$

The internal bias field was effectively reduced to zero within 40 s at 90 °C, while the time required at 30 °C was around 600 s. In contrast, both P_r and P_s continued to increase gradually over a period up to 3000 s at 30 °C, as shown in Fig. 5. Therefore, it is evident

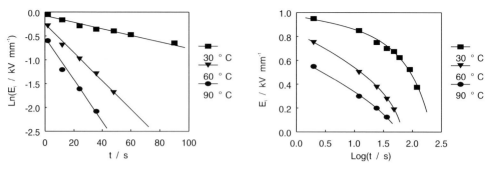

Fig. 4   Reduction in E_i as a function of time during field-driven deageing of unpoled materials at various temperatures (a) $\ln(E_i)$ vs t, (b) E_i vs log (t).

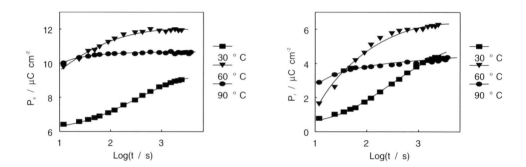

Fig. 5   Increase in (a) saturation and (b) remanent polarisation of unpoled materials during field-driven deageing at various temperatures.

that significant changes in the ferroelectric switching behaviour continued to take place long after the internal bias field had apparently been removed. Similar observations were made previously with regard to hard PZT ceramics.[10,11]

The initial P_r and P_s values obtained at 30 °C were significantly lower than those at 60 and 90 °C. This behaviour contradicts previous measurements taken on cooling from 150 °C under a continuous AC field, which showed gradual increases in both P_r and P_s with reducing temperature.[12] The most likely explanation for this difference in behaviour is that the domain wall stabilisation effects at 30 °C were so strong as to be effective at inhibiting domain wall motion even after continuous AC cycling for 2400 s. Effectively, this means that P_r and P_s did not reach their ultimate values, measured previously as 6.3 and 11.9 µC cm^{-2} respectively at 30 °C,[12] within the timescale of the experiment. This is an interesting observation, given the relatively short timescale over which the internal bias field was removed.

During ageing of the poled specimens at 30 and 60 °C, the P–E loops showed a gradual shift along the electric field axis associated with the build up of the internal bias field, as shown in Fig. 6(a). The value of E_i was higher at lower ageing temperatures, irrespective of the ageing time, as shown in Fig. 7. In earlier studies, the increase of E_i at lower ageing

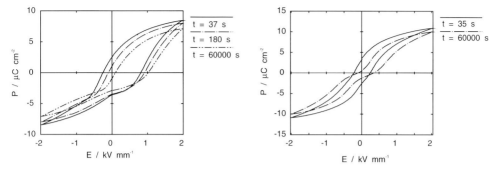

Fig. 6   Changes in P–E loops of poled materials during ageing at (a) 30 °C and (b) 90 °C.

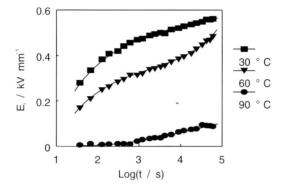

Fig. 7   Increase in E_i as a function of time during ageing of poled materials at various temperatures.

temperatures has been attributed to an increase in the depth of the potential well confining the oxygen vacancy, due to the reduction in dielectric permittivity.[6,7]

   The saturation polarisation P_s reduced significantly during ageing, the most pronounced change being found at the lower ageing temperatures, as shown in Fig. 8(a). The variation in remanent polarisation values during ageing was more complex, since the P–E loops developed a pronounced asymmetry, as shown in Figs 6 and 8. In order to fully describe this behaviour, it is necessary to report two P_r values, $P_r(1)$ and $P_r(2)$, corresponding to the 'upper' and 'lower' portions of the P–E loop respectively. The most significant of these is the lower value $P_r(2)$, since this represents the actual state of remanent polarisation in which the specimen was aged. Considering these $P_r(2)$ values, it is evident that the magnitude of P_r was initially around 2.1 μC cm^{-2} at 90 °C, increasing to 3.6 μC cm^{-2} at 30 °C and 5.0 μC cm^{-2} at 60 °C. The magnitude of P_r reduced gradually during ageing at each of these temperatures.

   An unusual ageing effect was observed at 90 °C, where the P–E loop of the poled specimen developed only a marked constriction (Fig. 6(b)), rather than the shift of the curves found at lower temperatures (Fig. 6(a)). This effect must be associated with the

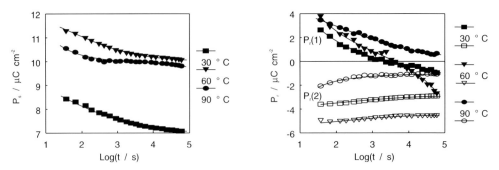

Fig. 8   Changes in (a) saturation and (b) remanent polarisation of poled materials during ageing at various temperatures.

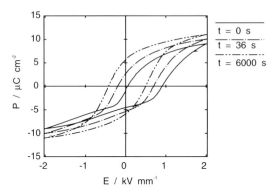

Fig. 9   Changes in P–E loops of poled materials during field-driven deageing at 30 °C ($E_0 = 2$ kV mm^{-1})

low remanent polarisation at 90 °C, due to the close proximity to the Curie point at 100 °C. The development of a constricted P–E loop can be understood in terms of the low P_r value, which implies a nearly random domain configuration comparable with that of the unpoled condition.

Field-driven deageing of the poled specimens caused a reversal of the ageing process, as illustrated by the P–E loops presented in Fig. 9. In this case, the reduction of the internal bias field during deageing did not follow the exponential time law, as shown in Fig. 10(a). Instead, the early deageing behaviour gave a reasonable fit to the logarithmic law reported previously for poled hard PZT[10] (Fig. 10(b)). However, the value of E_i remained relatively constant beyond 1000 s leaving a significant non-zero internal bias field ($E_i \approx 0.1$ kV mm^{-1}) at the end of the test procedure (after 12 000 s). This behaviour provides a contrast to that of the unpoled specimens (Fig. 4), in which the internal bias field was removed within 600 s, even at 30 °C. These results indicate the existence of a strong domain stabilisation process in the poled specimens, which was particularly resistant to field-forced deageing.

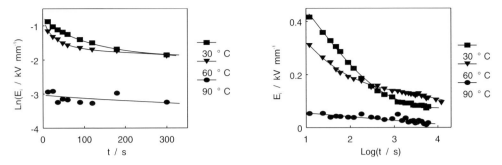

Fig. 10   Reduction in E_i as a function of time during field-driven deageing of poled materials at various temperatures (a) $\ln(E_i)$ vs t, (b) E_i vs $\log(t)$.

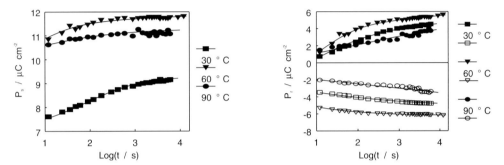

Fig. 11   Increase in (a) saturation and (b) remanent polarisation of poled materials during field-driven deageing at various temperatures.

The variations in P_r and P_s during deageing of the poled specimens also illustrate the manner in which the field-forced deageing procedure effectively reverses the ageing processes. The ultimate P_r and P_s values obtained at the end of the deageing procedure were in agreement with the initial values prior to the initiation of the ageing process (compare the curves presented in Figs 8 and 11).

4. CONCLUSIONS

Constricted P–E loops were obtained after ageing unpoled commercial cobalt-doped BaTiO$_3$ ceramics for 7 days at temperatures from 30 to 90 °C. The internal bias field in these unpoled specimens was reduced to zero within 600 s during field-forced deageing at 30 °C. Both P_r and P_s continued to increase over a longer timescale (up to 3000 s), indicating an improved degree of ferroelectric polarisation switching. Ageing of poled specimens at 30 and 60 °C gave rise to asymmetric P–E loops shifted along the electric field axis. Constricted loops were obtained at 90 °C as a result of a very low remanent polarisation in the vicinity of the Curie temperature ($T_c = 100$ °C).

The reduction of E_i during field-forced deageing of the unpoled specimens followed an exponential time law, whereas the poled specimens yielded a logarithmic decay. The internal bias field in such poled specimens proved to be extremely resistant to field-forced deageing, with the result that a significant non-zero value of E_i remained after continuous AC cycling for 12 000 s.

ACKNOWLEDGEMENTS

Mr. Ben-Omran would like to thank the Libyan Higher Education for financial support.

REFERENCES

1. A. Von Hippel: 'Ferroelectricity, domain structure, and phase transitions of barium titanate', *Rev. Mod. Phys.*, 1950, **22**, 221–237.
2. S.L. Swartz: 'Topics in electronic ceramics', *IEEE Trans. IE*, 1990, **25**, 935–986.
3. D. Berlincourt: 'Piezoelectric ceramic compositional development', *J. Acoust. Soc. Am.*, 1992, **91**, 3034–3040.
4. Morgan Matroc Ltd. (Unilator Division, Ruabon, UK): *Piezoelectric ceramic products*.
5. K. Carl and K.H. Hardtl: 'Electrical after-effects in $Pb(Ti,Zr)O_3$ ceramics', *Ferroelectrics*, 1978, **17**, 473–486.
6. R. Lohkamper, H. Neumann and G. Arlt: 'Internal bias in acceptor-doped $BaTiO_3$ ceramics : numerical evaluation of increase and decrease', *J. Appl. Phys.*, 1990, **68**, 4220–4224.
7. G. Arlt and H. Neumann: 'Internal bias in ferroelectric ceramics: origin and time dependence', *Ferroelectrics*, 1988, **87**, 109–120.
8. U. Robels and G. Arlt: 'Domain wall clamping in ferroelectrics by orientation of defects', *J. Appl. Phys.*, 1993, **73**, 3454–3460.
9. D.A. Hall, P.J. Stevenson and T.R. Mullins: 'Dielectric nonlinearity in hard piezo-electric ceramics', *Brit. Cer. Proc.*, 1997, **57**, 197–211.
10. D.A. Hall and P.J. Stevenson: 'Field induced destabilisation of hard PZT ceramics', *Ferroelectrics*, 1996, **187**, 23–37.
11. P.J. Stevenson and D.A. Hall: 'Ageing and field-forced deageing behaviour of hard PZT ceramics', *this Proceedings*.
12. D.A. Hall, P.J. Stevenson and M.M. Ben-Omran: 'Field and temperature dependence of dielectric properties in $BaTiO_3$ based piezoceramics', *J. Phys. Condensed Matter*, 1998, **10**, 461–476.

AGEING AND FIELD-FORCED DEAGEING BEHAVIOUR OF HARD PZT CERAMICS

P.J. STEVENSON AND D.A. HALL

Materials Science Centre, University of Manchester/UMIST, Grosvenor Street, Manchester M1 7HS, UK

ABSTRACT

The paper is concerned with the P–E (polarisation–electric field) switching behaviour of hard PZT (lead zirconate titanate) ceramics. The remanent polarisation P_r, saturation polarisation P_s, and internal bias field E_i were determined at a field amplitude of 3.5 kV mm^{-1} during ageing and field-driven deageing measurement procedures. It is shown that the rates of increase of P_r and P_r during field-driven deageing, which represents an AC poling process, are dependent both on temperature and on the prior ageing treatment of the unpoled specimens.

1. INTRODUCTION

Hard PZT (Lead Zirconate Titanate) ceramics are widely used in high power electromechanical transducers, for example in ultrasonic cleaners and SONAR transmitters.[1] The stability of such materials under high drive conditions, with field levels typically in the range 400 to 800 V mm^{-1},[2] is a direct result of ageing processes which act to restrict ferroelectric domain wall motion. In acceptor (e.g. Mn, Fe) doped perovskite ferroelectrics, the observed reductions in dielectric permittivity and loss during ageing are attributed to the presence of acceptor ion-oxygen vacancy defect associates.[3,4] These dipolar defects gradually reorient towards the direction of the local domain polarisation, resulting in a reduction of domain wall mobility.[5]

High field P–E (polarisation–electric field) hysteresis measurements indicate the presence of an effective internal field associated with these oriented defects, referred to as the 'internal bias field' E_i. For poled materials, the internal bias field is a measure of the relative difficulty in reorienting the polarisation away from the stabilised state i.e. $E_c(2) > E_c(1)$, as shown in Fig. 1(a). In unpoled materials, the domain stabilisation effects act to oppose polarisation changes in either sense, away from the randomly oriented domain structure. In this case, the observed constricted P–E loops can be understood in simple terms as a superposition of two P–E loops shifted towards the positive and negative sides of the electric field axis respectively.[3] The internal bias field in such a material can

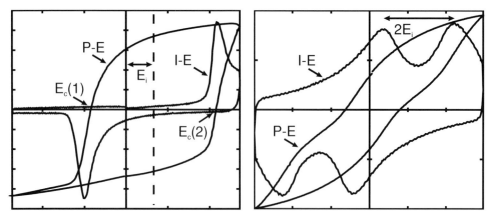

Fig. 1   Typical P–E and I–E loops for (a) poled and (b) unpoled acceptor-doped ferroelectrics after ageing.

be determined from the measured I–E (induced current-electric field) loop, as shown in Fig. 1(b).

The application of a continuous AC field at a level sufficiently high to induce polarisation switching causes a gradual randomisation in the orientations of the dipolar defects, resulting in a destabilisation of the domain structure and a reduction of the internal bias field. This phenomenon, which represents an effective reversal of the ageing effects, is usually referred to as 'hysteresis relaxation', the latter term having arisen initially to describe analogous effects in ferromagnetic materials.[3]

The ageing and deageing characteristics of hard PZT ceramics are of considerable practical significance, since they determine the ultimate dielectric and piezoelectric properties obtained (after ageing), as well as the stability of the material under high drive conditions. Furthermore, the restriction of ferroelectric domain wall motion in unpoled materials is likely to be an important factor in determining the efficiency and reproducibility of subsequent poling operations.

In a recent publication, the present authors described the field-forced deageing behaviour of a commercial hard PZT ceramic as functions of the drive conditions and temperature.[6] In that work, it was observed that the initial polarisation switching characteristics of an unpoled material were much less efficient than those of a poled material, at a given ageing time. For example, at a temperature of 90 °C the initial 'saturation' polarisation P_s for an unpoled material was 17 μC cm^{-2} whereas the equivalent value for a poled material was 33 μC cm^{-2}. This effect was attributed to the relative ease of 180 ° polarisation reversal in comparison with 90 ° domain switching. The apparent value of P_s increased for both poled and unpoled materials during field-forced deageing to give an ultimate value close to 40 μC cm^{-2}. The results obtained for the remanent polarisation P_r increased in a similar manner during the experiment. The time required to reach stable values of P_s and P_r is clearly of great interest with regard to the changes in the ferroelectric domain structure which occur during the poling operation.

The aim of the present study was to establish the influence of factors such as the time interval between a thermal deageing event (e.g. a high temperature electroding process) and the poling treatment on the subsequent poling behaviour of hard PZT ceramics. In the present work, a high AC electric field was employed in preference to the conventional DC poling field, since the former allows the determination of changes in parameters such as P_r, P_s and E_i during the poling procedure. The subsequent ageing characteristics of the poled materials were also investigated.

2. EXPERIMENTAL PROCEDURES

The material used in this study was a hard PZT ceramic, type PZ26, manufactured by Ferroperm Ltd (Hejreskovvej 6, DK–3490, Denmark). Specimens were provided in the form of poled, electroded disks with a diameter of 10 mm and a thickness of 1 mm. According to information provided by the manufacturer, the material was a Mn-doped PZT ceramic with a composition close to the morphotropic phase boundary. Earlier work demonstrated that the material had a tetragonally-distorted perovskite type crystal structure, with a c/a ratio at room temperature of 1.022, and a ferroelectric Curie point of 345 °C.[7]

High field measurements were conducted in a temperature-controlled silicone oil bath, which helped to prevent dielectric breakdown of the specimens and provided a stable temperature environment for the long-term ageing and deaging experiments. A triangular high voltage AC waveform was applied using a Thurlby Thandar TG1304 function generator in conjunction with a Chevin Research HVA1B ±5 kV amplifier. The induced current flowing through the specimen was converted to a voltage signal using a custom-built I–V converter, the resulting current and field waveforms being recorded on a PC via an Amplicon PC226 A/D card. P–E loops were obtained by numerical integration of the current with respect to time. Further details of the measurement system are given in a previous publication.[7] The amplitude of the electric field was chosen as 3.5 kV mm^{-1} for consistency with earlier studies.[6]

Prior to starting the main experiments, the specimens were thermally deaged/depoled by heating to a temperature of 450 °C for 15 minutes, after which they were aged at room temperature for various lengths of time, as required.

3. RESULTS AND DISCUSSION

3.1 DEAGEING BEHAVIOUR OF UNPOLED SPECIMENS

The first set of measurements was carried out to determine the influence of the time period between the thermal deageing treatment and the first application of the AC field on the subsequent field-forced deageing behaviour of unpoled, aged specimens. It was anticipated that the longer ageing times should give rise to a more stable ferroelectric domain structure, which would be more resistant to reorientation by the high AC field.

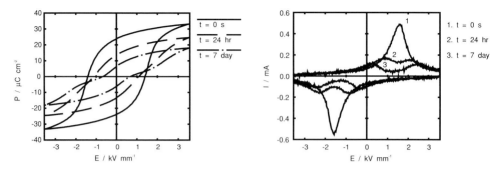

Fig. 2   Initial (a) P–E and (b) I–E hysteresis loops obtained for PZ26 ceramics at 90 °C, after various room temperature ageing schedules.

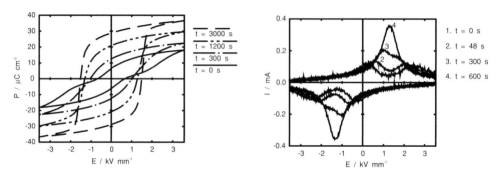

Fig. 3   Evolution of (a) P–E and (b) I–E loops for unpoled PZ26 ceramics during field-forced deageing at 90 °C, after prior ageing for 7 days at room temperature.

The initial P–E and I–E hysteresis loops obtained at 90 °C, after ageing for various time periods at room temperature, are illustrated in Fig. 2. Here, the influence of the ageing period can be seen in the progressively constricted P-E loops and the more pronounced splitting of the switching current peaks in the I–E curve which reflect an increased internal bias field. The initial values of P_s and P_r also reduced significantly as a result of ageing, from 33.1 and 23.7 μC cm^{-2} to 17.6 and 6.2 μC cm^{-2} respectively after ageing for 7 days.

The deageing characteristics of these specimens can be illustrated by the P–E and I–E loops obtained for a specimen aged for 7 days prior to the test, as shown in Fig. 3. It is evident that the application of a continuous high AC field effectively reversed the ageing process, as reported by previous authors,[3,6] removing the constriction from the P–E loop and yielding significantly higher P_s and P_r values.

An increase in the ageing time prior to the deageing experiment resulted in a progressive delay in the onset of enhanced polarisation switching, as shown by the results presented in Fig. 4(a). The rate of change of the internal bias field with time provided a reasonable fit to the exponential time law reported by Carl and Hardtl,[3] as shown in Fig. 4(b):

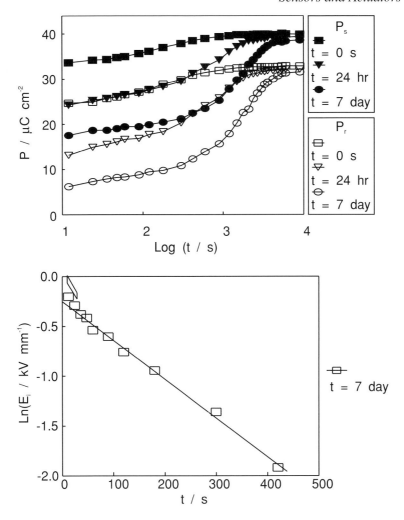

Fig. 4 Changes in (a) polarisation and (b) internal bias field for unpoled PZ26 ceramics during field-forced deageing at 90 °C, after various room temperature ageing schedules.

$$E_i(t) = E_i(0)e^{-\frac{t}{\tau}} \tag{1}$$

Here, t is the deageing time and τ is a characteristic time constant or 'relaxation time'.

A series of experiments was conducted to determine the influence of temperature on the deageing behaviour. In this case, a set of specimens was thermally deaged and then aged in the unpoled state for a period of 7 days. During subsequent field-forced deageing, it was evident that the enhancement of polarisation switching was delayed significantly at lower temperatures, as shown in Fig. 5(a). For example, a P_r value of 30 μC cm^{-2} was

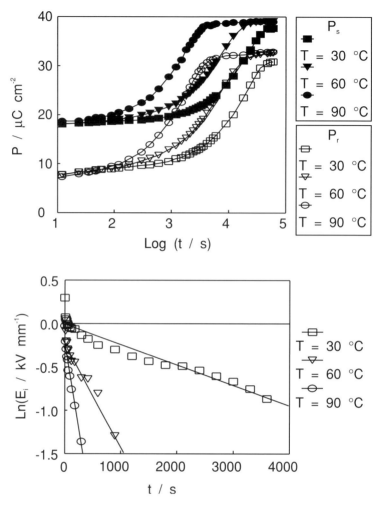

Fig. 5   Changes in (a) polarisation and (b) internal bias field for unpoled PZ26 ceramics during field-forced deageing at various temperatures, after prior ageing for 7 days at room temperature.

obtained after deageing for 3600 s at 90 °C, whereas 48 000 s were required to reach the same level at 30 °C. The ultimate values of both P_r and P_s were relatively unaffected by changes in temperature.

The thermally-activated nature of the deageing process was evident in an increased rate of reduction of E_i with time at higher temperatures, as shown by the results presented in Fig. 5(b). Previous authors[3,4] described the effect of temperature on field-forced deageing in terms of a thermally-activated relaxation time τ:[3,4]

$$\tau(T) = \tau_\infty e^{\frac{W}{kT}} \tag{2}$$

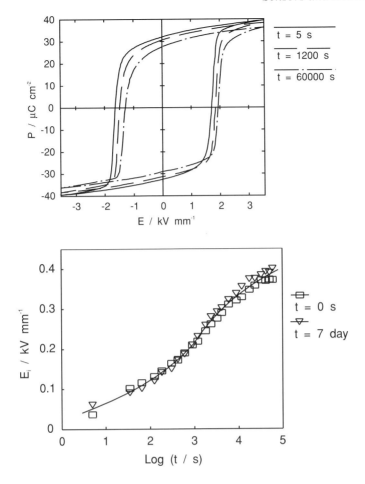

Fig. 6 (a) Evolution of P–E loops and (b) build up of internal bias field during ageing of poled PZ26 ceramics at 90 °C, after various room temperature ageing schedules and subsequent field-forced deageing.

The results obtained in the present study (Fig. 5) yield τ values of 5300, 1100, and 250 s at temperatures of 30, 60, and 90 °C respectively, indicating an activation energy W of approximately 0.47 eV. This value is comparable with those reported by Carl and Hardtl for similar acceptor-doped PZT ceramics.[3] It was recognised by Neumann and Arlt that this activation energy is related to the defect dipole reorientation process, and therefore is dependent on factors such as the depth of the potential well associated with the position of the oxygen vacancy in the perovskite unit cell.[4]

3.2 SUBSEQUENT AGEING OF POLED SPECIMENS

Immediately after completion of the field-forced deageing procedures, two complete

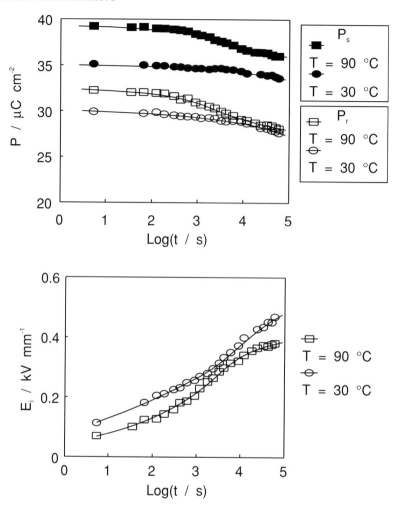

Fig. 7 Changes in (a) polarisation and (b) internal bias field during ageing of poled PZ26 ceramics at various temperatures.

voltage cycles were applied to the specimens to induce a state of remanent polarisation. Hysteresis measurements were then conducted by applying two voltage cycles at selected ageing times. During ageing, the P–E loops showed a gradual shift towards the positive side of the electric field axis, indicating a gradual build up of the internal bias field. A typical set of data is presented in Fig. 6(a) to illustrate this effect.

Specimens subjected to different ageing treatments (in the unpoled state) prior to the field-driven deageing test exhibited very similar ageing characteristics in the poled condition, as shown in fig. 6(b). The similarity in the ageing behaviour of these poled specimens was anticipated, since the P_r values obtained on completion of the deageing test were almost identical, as described above (Fig. 4(a)). The general form of the E_i vs t

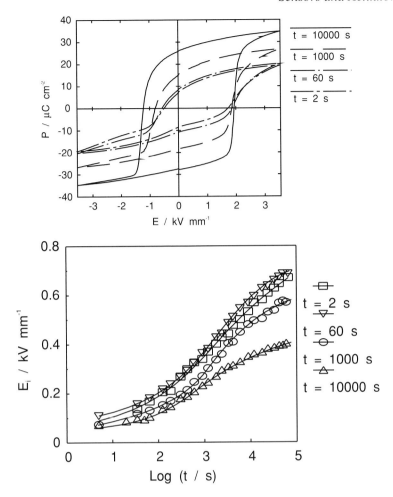

Fig. 8 (a) P–E loops (at t = 6 x 10⁴ s) and (b) build up of internal bias field during ageing of PZ26 ceramics subjected to various pre-cycling treatments at 90 °C.

relationship was logarithmic, although there was an increase in the gradient of the curve after 100 s, and the results showed a tendency towards a constant saturation value at long ageing times. In this respect, the results were comparable with those reported by Arlt and Neumann for Ni-doped barium titanate ceramics.[5]

The effect of temperature on the ageing behaviour of poled specimens is illustrated by the results presented in Fig. 7. The rate of increase of E_i appeared to be independent of temperature at ageing times less than 1000 s. Beyond this time, an acceleration in the ageing rate and then a tendency towards a saturated level was increasingly evident at higher temperatures. The E_i results obtained at 30 °C showed no indication of saturation within the time scale of the experiments (up to 6 x 10⁴ s).

A final series of experiments was conducted to determine the influence of the initial remanent polarisation on the subsequent ageing behaviour. Freshly thermally-deaged specimens were subjected to various pre-cycling treatments, from 2 to 10 000 s, prior to initiation of the ageing experiment. This procedure resulted in a gradual increase in P_r with pre-cycling time, yielding values of 7, 8, 15 and 26 µC cm^{-2} for pre-cycling times of 2, 60, 1000 and 10 000 s respectively, as shown in Fig. 8(a). The effect of a continuous AC field on the polarisation switching behaviour of such thermally deaged specimens was described in more detail in a previous publication.[8]

During ageing, all of the P–E loops exhibited a shift towards the positive side of the electric field axis, corresponding to an increased internal bias field, as shown in Fig. 8(a). Somewhat surprisingly, the less efficiently poled specimens, which had been subjected to the shortest pre-cycling schedules, consistently exhibited the highest E_i values. For example, ultimate E_i values of 0.67, 0.69, 0.57 and 0.40 kV mm^{-1}. were achieved for pre-cycling times of 2, 60, 1000 and 10 000 s respectively (Fig. 8(b)). Robels *et al.* recognised that the measured macroscopic value of E_i depends principally on the extent of 180 $^\circ$ domain reversal, relative to the local direction of E_i, during the measurement procedure. Reorientation of the polarisation into directions at 90 $^\circ$ to the local direction of E_i is effectively symmetric with respect to the sense of the applied field, and therefore should not contribute to a shift of the P–E loop along the electric field axis.[9] The present authors have previously demonstrated how the apparent value of E_i depends upon the amplitude of the measuring field, which also can be related to the extent of 180 $^\circ$ domain switching induced by the applied AC electric field.[6]

Therefore, the results presented in Fig. 8(b), which show the build up of E_i during ageing of these specimens, appear to suggest that the degree of 180 $^\circ$ domain switching was greater for specimens which had been subjected to shorter pre-cycling treatments. However, this proposal is contradicted by the lower P_s values of those specimens, which are evident in Fig. 8(a). It is not possible to develop this argument further at present, in the absence of more detailed information on the relative contributions of 90 $^\circ$ and 180 $^\circ$ domain switching to P_s.

On inspection of the results presented in Fig. 8(a), it is evident that the P–E loops obtained using the shorter pre-cycling times, which yielded the higher E_i values, exhibited a pronounced constriction as well as a shift along the electric field axis. This distortion of the loop, which shows some similarity to the P–E characteristic of an unpoled aged material (as shown in Fig. 1(b)), reflects the low P_r values obtained in these specimens and could also have exerted some influence on the measured internal bias field.

It is evident from the results presented in Figs 4 and 5 that the time required to reach stable values of P_r and P_s under a continuous high AC electric field are dependent both on temperature and on the ageing time (in the unpoled condition) prior to application of the field. At a temperature of 90 °C, a time period of approximately 3 hours under a continuous AC field of 3.5 kV mm^{-1} was required to achieve 'saturated' values of P_s and P_r (Fig. 5). The influence of these variables on the field-forced deageing process is clearly relevant if a high P_r value is to be obtained during poling. It is anticipated that a significantly shorter deageing time would be necessary at the higher temperatures (typically ~ 130 °C) which are commonly employed for poling hard PZT ceramics. Therefore, the use of elevated temperatures during poling serves the dual purpose of reducing the coercive field and reducing the time required for field-forced deageing, both of which help to achieve a high degree of remanent polarisation.

4. CONCLUSIONS

The initial P_r and P_s values obtained on the application of a high AC electric field (E_0 = 3.5 kV mm^{-1}) reduced significantly as a result of prior ageing (in the unpoled condition) at room temperature. Both P_r and P_s were found to increase during the field-driven deageing procedure; the time required to reach 'saturation' of these values increased with a reduction in temperature and with an increase in the prior ageing time. The ageing characteristics of specimens subjected to different pre-cycling schedules showed that the apparent internal bias field E_i was higher for specimens with lower P_r values, which appeared to contradict the view that higher E_i values should be obtained by more efficient polarisation switching during the measurement procedure.

ACKNOWLEDGEMENTS

The authors wish to thank DRA(Holton Heath) for financial support and Ferroperm Ltd. for supplying the PZT test specimens.

REFERENCES

1. D. Berlincourt: 'Piezoelectric ceramic compositional development', *J. Acoust. Soc. Am.*, 1992, **91**, 3034–3040.
2. J.L. Butler, K.D. Rolt and F. Tito: 'Piezoelectric ceramic mechanical and electrical stress study', *J. Acoust. Soc. Am.*, 1994, **96**, 1914–1917.
3. K. Carl and K.H. Hardtl: 'Electrical after-effects in Pb(Ti,Zr)O$_3$ ceramics', *Ferroelectrics*, 1978, **17**, 473–486.
4. H. Neumann and G. Arlt: 'Dipole orientation in Cr-modified BaTiO$_3$ ceramics', *Ferroelectrics*, 1987, **76**, 303–310.
5. G. Arlt and H. Neumann: 'Internal bias in ferroelectric ceramics: origin and time dependence', *Ferroelectrics*, 1988, **87**, 109–120.
6. D.A. Hall and P.J. Stevenson: 'Field induced destabilisation of hard PZT ceramics', *Ferroelectrics*, 1996, **187**, 23–37.
7. D.A. Hall, P.J. Stevenson and T.R. Mullins: 'Dielectric nonlinearity in hard piezo electric ceramics', *Brit. Cer. Proc.*, 1997, **57**, 197–211.
8. D.A. Hall and C.E. Millar: 'High field de-ageing and fatigue characteristics of hard PZT ceramics', *Proc. Electroceramics IV*, 1994, 417–420.
9. U. Robels, A. Mellage and G. Arlt: 'Ageing after polarisation reversal in acceptor-doped ferroelectrics', *Proc. Electroceramics IV*, 1994, 373–376.

INDEX